Basic Statistics in Multivariate Analysis

POCKET GUIDES TO
SOCIAL WORK RESEARCH METHODS

Series Editor
Tony Tripodi, DSW
Professor Emeritus, Ohio State University

Determining Sample Size: Balancing Power, Precision, and Practicality
Patrick Dattalo

Preparing Research Articles
Bruce A. Thyer

Systematic Reviews and Meta-Analysis
Julia H. Littell, Jacqueline Corcoran, and Vijayan Pillai

Historical Research
Elizabeth Ann Danto

Confirmatory Factor Analysis
Donna Harrington

Randomized Controlled Trials: Design and Implementation for Community-Based Psychosocial Interventions
Phyllis Solomon, Mary M. Cavanaugh, and Jeffrey Draine

Needs Assessment
David Royse, Michele Staton-Tindall, Karen Badger, and J. Matthew Webster

Multiple Regression with Discrete Dependent Variables
John G. Orme and Terri Combs-Orme

Developing Cross-Cultural Measurement
Thanh V. Tran

Intervention Research:

Developing Social Programs
Mark W. Fraser, Jack M. Richman, Maeda J. Galinsky, and Steven H. Day

Developing and Validating Rapid Assessment Instruments
Neil Abell, David W. Springer, and Akihito Kamata

Clinical Data-Mining: Integrating Practice and Research
Irwin Epstein

Strategies to Approximate Random Sampling and Assignment
Patrick Dattalo

Analyzing Single System Design Data
William R. Nugent

Survival Analysis
Shenyang Guo

The Dissertation: From Beginning to End
Peter Lyons and Howard J. Doueck

Cross-Cultural Research
Jorge Delva, Paula Allen-Meares, and Sandra L. Momper

Secondary Data Analysis
Thomas P. Vartanian

Narrative Inquiry
Kathleen Wells

Structural Equation Modeling
Natasha K. Bowen and Shenyang Guo

Finding and Evaluating Evidence: Systematic Reviews and Evidence-Based Practice
Denise E. Bronson and Tamara S. Davis

Policy Creation and Evaluation: Understanding Welfare Reform in the United States
Richard Hoefer

Grounded Theory
Julianne S. Oktay

Systematic Synthesis of Qualitative Research
Michael Saini and Aron Shlonsky

Quasi-Experimental Research Designs
Bruce A. Thyer

Conducting Research in Juvenile and Criminal Justice Settings
Michael G. Vaughn, Carrie Pettus-Davis, and Jeffrey J. Shook

Qualitative Methods for Practice Research
Jeffrey Longhofer, Jerry Floersch, and Janet Hoy

Analysis of Multiple Dependent Variables
Patrick Dattalo

Culturally Competent Research: Using Ethnography as a Meta-Framework
Mo Yee Lee and Amy Zaharlick

Using Complexity Theory for Research and Program Evaluation
Michael Wolf-Branigin

Basic Statistics in Multivariate Analysis
Karen A. Randolph and Laura L. Myers

KAREN A. RANDOLPH
LAURA L. MYERS

Basic Statistics in Multivariate Analysis

OXFORD
UNIVERSITY PRESS

OXFORD
UNIVERSITY PRESS

Oxford University Press is a department of the University of Oxford.
It furthers the University's objective of excellence in research, scholarship,
and education by publishing worldwide.

Oxford New York
Auckland Cape Town Dar es Salaam Hong Kong Karachi
Kuala Lumpur Madrid Melbourne Mexico City Nairobi
New Delhi Shanghai Taipei Toronto

With offices in
Argentina Austria Brazil Chile Czech Republic France Greece
Guatemala Hungary Italy Japan Poland Portugal Singapore
South Korea Switzerland Thailand Turkey Ukraine Vietnam

Published in the United States of America by
Oxford University Press
198 Madison Avenue, New York, NY 10016

Library of Congress Cataloging-in-Publication Data

Randolph, Karen A.
Basic statistics in multivariate analysis / Karen A. Randolph, Laura L. Myers.
p. cm. — (Pocket guides to social work research methods)
Includes bibliographical references and index.
ISBN 978–0–19–976404–4 (pbk. : alk. paper) 1. Social service—Research—
Methodology. 2. Multivariate analysis. I. Myers, Laura L. II. Title.
HV11.R3123 2013
519.5′35—dc23
2012033754

Contents

Acknowledgments vii

1 Introduction 1

2 Descriptive Statistical Methods 11

3 Inferential Statistics 35

4 Bivariate Statistical Methods 69

5 Bivariate and Multivariate Linear Regression Analysis 109

6 Analysis of Variance (ANOVA) and Covariance (ANCOVA) 133

7 Path Analysis 163

Appendix: Statistical Symbols 187

Glossary 189

References 203

Index 209

Acknowledgments

We are grateful for the generous encouragement and helpful advice from a number of people who contributed to the preparation of this book. First and foremost, we thank Dr. Tony Tripodi, the series editor of *Pocket Guides to Social Work Research Methods*, for his unwavering support and enthusiasm. We are also very appreciative of the assistance we received from Maura Roessner, Editor, and Nicholas Liu, Assistant Editor, at Oxford University Press. Karen Randolph is grateful for the support of Mrs. Agnes Flaherty Stoops through the Agnes Flaherty Stoops Professorship in Child Welfare. We have been fortunate to work with many bright and talented students, whose commitment to understanding and applying complex statistical methods in conducting their own social work–based research inspired us to pursue this topic. A very special acknowledgment is extended to Christina Ouma, doctoral student and contributing author of this book's companion website, for her tireless efforts and attention to detail in navigating the National Educational Longitudinal Study of 1988 to develop the practice exercises for the statistical procedures. We also thank Leah Cheatham, Hyejin Kim, Dr. Bruce Thyer, and David Albright for their thoughtful comments on manuscript drafts. Karen Randolph would like to thank Dr. Betsy Becker, Ying Zhang, Leah Cheatham, and Tammy Bradford for their assistance on Chapter Six.

Basic Statistics in Multivariate Analysis

1

Introduction

Statistical methods used to investigate questions that are relevant to social work researchers are becoming more complex. The use of methods such as path analysis in causal modeling is increasingly required to match appropriate data analysis procedures to questions of interest. As a consequence, social work researchers need a skill set that allows them to thoroughly understand and test multivariate models accurately. A strong background in basic statistics provides the foundation for this skill set and allows for the use of more advanced methods to study relevant questions to social work researchers.

The purpose of *Basic Statistics in Multivariate Analysis* is to introduce readers to three multivariate analytical methods, with a focus on the basic statistics (e.g., mean, variance) that support these methods. Multivariate analytical methods are made up of basic statistical procedures. This is an important, yet often overlooked, aspect of advanced statistics. We posit that, by having a strong foundation in basic statistics, particularly with regard to understanding their role in more advanced methods, readers will be more confident and thus more likely to utilize advanced methods in their research.

What do we mean by "basic statistics"? Basic statistics are statistics that organize and summarize data. This includes frequency distributions, percentages, measures of central tendency (i.e., mean, median, mode), and measures of dispersion or variability (i.e., range, variance, and standard deviation). Basic statistics are also referred to as descriptive statistics (e.g., Rubin, 2010), as the intent is to describe individual variables, rather than test inferences about the relationships between variables. The results of basic statistical analysis, also called univariate analysis, are often displayed in charts and graphs, such as bar graphs, histograms, and stem and leaf plots.

What do we mean by "multivariate analysis"? The focus of multivariate analysis methods is on multiple variables. It is a collection of statistical techniques that is used to examine and make inferences about the relationships between variables. Real world problems that are of interest to social workers are generally affected by a variety of factors. Multivariate analysis allows social work researchers to understand the etiology of these problems in a way that more accurately reflects how they really happen. We can examine the relationships between several factors (i.e., variables) and an outcome by using multivariate analytic methods. While several multivariate analytic methods are available, we describe three of the more common methods—multiple linear regression analysis, analysis of variance (ANOVA) and covariance (ANCOVA), and path analysis. More information about each of these methods is provided later in this chapter.

Bivariate statistics make up a special class of multivariate statistics. As the name implies, bivariate statistics focus on the relationship between two variables. Common bivariate statistical tests are the dependent samples *t*-test, the independent samples *t*-test, the Pearson *r* correlation, and the chi-square test. These tests, and others, are used to test inferences about the relationship between two variables.

In general, books on basic statistics for social workers address an important educational need in social work education—to facilitate the development of skills for *consuming* information about statistics. This is based on an assumption that the target audience is unlikely to extend their studies in a way that includes the need to *learn and conduct multivariate analysis*. Our focus is different. We take a unique approach by directing our efforts toward preparing entry-level doctoral students and early-career social work researchers, especially those who may not have a strong background in basic statistics, to use advanced analytic procedures by highlighting the important role of basic statistics in these methods. In their content review of 30 statistical textbooks, Hulsizer and

Woolf (2009) observed that only "a small handful of authors elected to go beyond simple regression and include a chapter on multiple regression (13%)" (p. 35). They also note the absence of content on Analysis of Covariance (ANCOVA) and other multivariate methods. We include content beyond simple regression to address these gaps.

The primary focus of this book is to offer opportunities for readers, particularly entry-level doctoral students and early-career social work researchers, to strengthen their understanding and skills in basic statistics and related statistical procedures so that they are more prepared to utilize multivariate analytical methods to study problems and issues that are of concern to social workers. We assume that readers have familiarity with univariate and bivariate statistical analysis and some experience in using the Statistical Package for the Social Sciences (SPSS) and AMOS software (SPSS Inc., 2011). The book is also designed to be used as a reference guide in addressing questions that may emerge in conducting multivariate analysis, as well as a companion text in advanced statistics courses for doctoral students.

THE BRIDGE FROM BASIC TO INFERENTIAL STATISTICS IN DATA ANALYSIS

This book provides information about both basic and inferential statistics. Basic statistics summarize or classify the characteristics of a sample. They provide a foundation for understanding the sample. For example, basic statistics can be used to indicate the number or percentage of males and females in a study, their mean or average age, and the range of their ages from youngest to oldest. Basic statistics include counts, percentages, frequency distributions, measures of central tendency, and measures of variability. They can be displayed as various graphical representations of the data.

While basic statistics provide information about a sample, inferential statistics focus on the population from which the sample was drawn, using data collected from the sample. Inferential statistics are used to make predictions or draw conclusions about the population based on what is known about the sample. Probability theory provides the basis for making predictions about a population from a sample. Inferential statistics include parametric statistical tests such as the Pearson's *r* correlation, Student's *t*-tests, and analysis of variance, and nonparametric statistical tests such as Spearman's rho, Mann–Whitney *U*, Wilcoxson signed rank, and Kruskal–Wallis *H* tests. As an example, the Pearson *r* correlation test could be used to determine the relationship between depression and frequency of alcohol use among older adults.

Basic and inferential statistics differ based on what their intended purpose is with regard to the type of information they provide. Basic statistics are used to summarize information about a sample. Inferential statistics are used to make predictions about a population based on information obtained from a sample of the population.

The process of making predictions from a sample to a population with inferential statistics is more restrictive than summarizing data using basic statistics. Because inferential statistics involve statistical testing, four assumptions about the data must be met. In general, the assumptions are as follows: 1) the dependent variable (DV) is measured at the interval or ratio level, 2) the distribution of the data is normal (i.e., unimodal and not excessively skewed or kurtotic), 3) the variances across the data are equal (i.e., homogeneity of variance), and 4) the observations are independent of one another. Note that, for some tests (e.g., dependent-samples *t* test), the 4th assumption (i.e., independence of observations) does not apply. This is the case when data are collected from the same sample at more than one time point (e.g., pre- and posttest observations). Furthermore, the manner in which some assumptions are operationalized varies depending on the particular type of parametric test. Finally, additional assumptions are required for tests of multivariate models when causality is inferred. All of this can be confusing. We will cover assumptions in much more detail throughout the book, including how to determine whether assumptions have been met and the impact on results when assumptions are violated.

Making predictions using inferential statistical tests also requires that models are accurately estimated. The following criteria are used to ensure accuracy in model estimation:

- The model should be correctly specified. A correctly specified model is one in which 1) all relevant independent variables (IV) are in the model, 2) all irrelevant IVs are not in the model, 3) each IV is measured without error, and 4) the IVs in the model are not correlated with variables that are not in the model.

- The IVs should not be strongly correlated with one another (i.e., no undue multicollinearity).
- There should be no influential outliers among the IVs or in the solution.
- The sample size should be large enough to detect results at the desired effect size.

We will also discuss these criteria in more detail in subsequent chapters, including how to determine whether each criterion has been met and when a criterion is not met, the extent to which it becomes problematic in model testing.

Basic and inferential statistics are related to one another in that basic statistics provide the foundation for conducting multivariate analyses, in order to make inferences about the relationship between variables. Kleinbaum and others (1988) describe this succinctly:

> The primary goal of most statistical analysis is to make statistical inferences, that is, to draw valid conclusions about a population of items of measurements based upon information contained in a sample from that population. Once sample data have been collected, it is useful, prior to analysis, to examine the data using tables, graphs, and [basic] statistics, such as the sample mean or the sample variance. Such descriptive efforts are important for representing the essential features of the data in easily interpretable terms. Following such examination, statistical inferences are made through two related activities: estimation and hypothesis testing. (p. 16)

AN INTRODUCTION TO MULTIVARIATE ANALYSIS IN SOCIAL WORK

In this book we describe how basic statistics are used to inform three common multivariate analytical methods—multiple linear regression analyses, analysis of variance (ANOVA) and covariance (ANCOVA), and path analysis. Often these methods are used to support making inferences about causality between variables. Of course, inferring causality requires more than just establishing a statistical association between variables. Other conditions are 1) the presumed cause (e.g., X) occurs before the presumed effect (e.g., Y) (i.e., time precedence), 2) the direction of the causal relationship (e.g., X causes Y rather than the other way around) is correctly specified (i.e., correct effect priority), and 3) there are no other plausible explanations of the relationship between the presumed cause and the presumed effect (i.e., nonspuriousness) (Kline, 2011, p. 98). We will revisit the conditions for establishing causality, particularly with regard to time order and nonspuriousness as these criteria are important in path analysis, discussed in Chapter 7.

Note also that each of these methods is a form of the general linear model. The basis of the general linear model is that "relationships among dependent and independent variables vary according to straight-line patterns" (Bohrnstedt & Knoke, 1994, p. 24). This is reflected in the way relationships between variables are typically hypothesized to vary together—e.g., the greater the change in X, the greater the change in Y. In fact, this is an additional assumption in multivariate analysis. In the next section, we provide an introduction to each method.

Multiple linear regression analysis is used to examine relationships between two or more IVs and one DV, measured on an interval or ratio scale. For example, a researcher could use linear regression analysis to identify factors that predict depression among recently unemployed factory workers. In this case, the DV is depression. It could be assessed using the Center for Epidemiologic Studies Depression Scale (CES-D), which is a 20-item inventory that provides a measure of depression as captured by scores that range between 0 and 60 (CES-D, Radloff, 1977). The CES-D measures depression at the ratio level. Possible predictors of depression among recently unemployed factory workers might be the number of months unemployed and the level of social support. These predictors can be included in the analysis as IVs. Multiple linear regression analysis allows us to determine which of these IVs is related to depression among recently unemployed factory workers. *Linear* regression analysis is differentiated from another type of multiple regression analysis, called *logistic* regression analysis. Logistic regression analysis is used when the DV is dichotomous, with only two values. The focus in this book is on linear regression analysis in which the DV is measured on an interval or ratio-level scale.

When we wish to examine average differences between two or more groups on factors that are measured on a nominal scale, we can use analysis of variance (ANOVA) statistical procedures. Because ANOVA is used to determine **average** (i.e., mean) differences, the DV must be numeric and measured at the interval or ratio level. ANOVA procedures are often used in experimental research to examine mean differences between treatment and control groups. For example, ANOVA procedures could be used to test an intervention designed to increase awareness of sexually transmitted diseases (STD) among young adults, as was the case in a study conducted by Garcia-Retamero and Cokely (2011). The DV is STD awareness and the IV is an indicator of group membership—the intervention or control group. ANOVA could be used to determine mean differences on STD awareness at posttest.

There are several statistical procedures within the ANOVA family (i.e., one-way ANOVA, two-way ANOVA, factorial ANOVA, multivariate ANOVA [MANOVA], and ANCOVA [analysis of covariance]). In this book we describe one-way ANOVA, two-way ANOVA, and ANCOVA procedures. One-way ANOVA is used to compare means across multiple groups. One-way ANOVA models include one IV measured at the nominal level and one DV measured at the interval or ratio level. Two-way ANOVA is used to compare means in models with two nominal-level IVs and one interval or ratio-level DV. ANCOVA is an extension of one- and two-way ANOVA. ANCOVA models include one or more nominal-level IVs, one interval or ratio-level DV, and an additional variable measured at the interval or ratio level, referred to as a covariate. The covariate is posited to influence the DV, in addition to the IV. Thus, the inclusion of the covariate is used to explain part of the variance of the DV, beyond what the IV explains.

The set of procedures available in the ANOVA family is actually a subset of multiple linear regression analysis procedures. While both sets of procedures are used to explain variation in an outcome (the DV) through variation in other factors, such as the treatment condition examined in an ANOVA model or the IVs in a multiple linear regression model, ANOVA models include only nominal-level IVs whereas multiple linear regression analysis allows for IVs that are measured at any level.

Path analysis is similar to multiple linear regression analysis in that it is used to examine linear relationships between variables. As such, path models include DVs that are measured on an interval or ratio-level scale. One major difference is that path analysis techniques allow us to more fully examine the nature of relationships between IVs and the DV. The results provide more specific information about the direct and indirect effects of IVs on the DV. This information provides a stronger basis for claims about the causal relationship between the variables. For example, Azuma and Chasnoff (1993) used path analytic methods to investigate the direct and indirect effects of prenatal drug exposure (the IV) on cognitive functioning (the DV) of a group of 3-year-old children. Findings showed that prenatal drug exposure had a direct negative effect on cognitive functioning at 3 years of age. Findings also showed that prenatal drug exposure affected cognitive functioning indirectly through children's poor perseverance and externalizing behaviors. These results show the nature of the relationship between prenatal drug exposure and children's subsequent cognitive functioning in greater detail.

ORGANIZATION OF THE BOOK

The information in this book is presented in seven chapters, including this introductory chapter. Chapters 2, 3, and 4 provide a review of basic statistics, hypothesis testing with inferential statistics, and bivariate statistical analytic methods, respectively. Chapter 5 describes bivarate and multiple linear regression analyses, Chapter 6 covers analysis of variance (ANOVA) and covariance (ANCOVA), and Chapter 7 focuses on path analysis, including a discussion of how basic statistics inform these methods. We also provide a key of statistics symbols and a glossary with definitions of important terms that are used throughout the book.

In each chapter, we introduce readers to the various basic statistical procedures by providing brief definitions, mathematical formulas, descriptions of the underlying logic and assumptions of each procedure, and examples of how they have been used in the social work research literature, particularly with diverse populations. We also provide brief step-by-step directions for conducting these procedures in SPSS and AMOS (SPSS, Inc. 2011). At the end of Chapters 5, 6, 7, we offer a list for further readings. Finally, the book offers a companion website (www.oup.com/us/pocketguides/) that provides detailed instructions for conducting each procedure, as well as data sets and worked examples based on data from the National Educational Longitudinal Study of 1988 (NELS: 88).

AN OVERVIEW OF THE NATIONAL EDUCATIONAL LONGITUDINAL STUDY OF 1988

The book's practice examples are based on data from the National Educational Longitudinal Study of 1988 (NELS: 88). NELS: 88 is a longitudinal study of the educational experiences of a nationally representative sample of youth that began in 1988 when these youth were in the eighth grade. The purpose of the study is to provide information about "critical transitions experienced by students as they leave middle or junior high school, and progress through high school and into postsecondary institutions or the work force" (United States Department of Education, National Educational Longitudinal Study of 1988, 2011, "Overview," para. 4). Data were collected from a nationally representative sample of eighth graders, their parents, teachers, and high school principals, and was supplemented with information from students' existing school records. Participants were first surveyed in 1988 and then again in 1990, 1992, and 2000. The database includes information on a wealth of topics such as "school, work, and home experiences; educational resources and support; the role of parents and peers in education; self-reports on smoking, alcohol and drug use, and extracurricular activities; and results of achievement tests in reading, social studies, mathematics and science" (United States Department of Education, National Educational Longitudinal Study of 1988, 2011, "Overview," para. 1). We use data from the first and second waves of NELS: 88 to demonstrate the statistical techniques described in the book.

2

Descriptive Statistical Methods

Descriptive statistical methods are used to summarize all of the data in an existing database into fewer numbers, making the data easier to visualize and understand. Faulkner and Faulkner (2009) define **descriptive statistical methods** as "ways of organizing, describing, and presenting quantitative (numerical) data in a manner that is concise, manageable, and understandable" (p. 155). Descriptive statistics utilize **univariate statistical methods** to examine and summarize data one variable at a time. We can calculate numeric values that describe samples or populations. Numeric values that describe samples are called **statistics**, whereas numeric values that describe populations are called **parameters**. This chapter focuses on a review of the descriptive statistical methods commonly used in social work research. Before we turn to these individual methods, we will first look at the steps involved in defining the variables that will be used in a study, and in determining how and at what level these variables will be measured.

DEFINING VARIABLES

Variables are concepts or characteristics that vary. **Constants** are those concepts in a research study that do not vary. For example, suppose we are trying to determine depression levels of a group of sixth grade girls. The concept in this study that will vary (variable) is the depression level, and two concepts or characteristics that do not vary (constants) are gender (girls) and grade in school (sixth). The process of defining a variable is called **conceptualization**. For example, in the previous example, we would first have to define what we mean by depression. Some people define depression based on the presence of negative emotions, while others define it as a series of behavioral symptoms. Still others view it as a combination of both emotions and behaviors. Some may prefer to ask the participants in the study to keep a log of how often they feel depressed, the duration of each depressive episode, and at what level of intensity the depression is experienced.

After a variable has been conceptualized, the next step for the researcher is to determine how the variable will be measured. This is called **operationalization.** Of course, how the variable is conceptualized affects how it will be measured or operationalized. There are standardized instruments, such as the *Brief Depression Rating Scale,* that measure the presence of depressive emotions, such as despair and anxiety, as well as behavioral symptoms that have been tied to depression, such as sleep disturbances and suicidal behaviors (Kellner, 1986). The researcher may create a log that the participants can make an entry in every time they experience depression, noting time, length, and intensity. Researchers generally turn to the social science research literature for assistance in conceptualizing and operationalizing variables. Many concepts of interest to social workers have been defined and measured many times by researchers. Often, these previously defined variables and measures can be adapted for use in new research studies.

Values and Value Categories

The way we operationalize the variables of interest in our research study determines the possible values our variable can take. For example, if we measure loneliness using a self-report scale from 0 (not lonely at all) to 10 (lonely most or all of the time), then the variable "loneliness" can take the values 0 to 10. If we define it as the number of times the client reports feeling lonely during a one-week period, then the variable can be equal to 0 and greater. If we measure it using the *UCLA Loneliness Scale* (Russell, 1996), then it can have any value from 20 to 80 (the possible values of this scale).

Variables can be classified as continuous or discrete depending on how they are operationalized (i.e., the set of values that they can assume). A **continuous variable** is a variable that can, in theory, take on an infinite value at any decimal point in a given interval. Examples of continuous variables are temperature, height, and length of time. Of course, while these variables can theoretically take on an infinite number of values, they are actually measured discretely, on a fixed number of decimal points. For example, while theoretically temperature is a continuous variable, we may choose to measure it to the nearest degree. A **discrete variable** can only take on a finite value, typically reflected as a whole number. Examples of discrete variables are grades on a final exam, the number of children in a family, and annual income. A discrete variable that assumes only two values is called a **dichotomous variable.** A variable that designates whether a person is in the experimental or control group is an example of a dichotomous variable. This variable would have two values—assignment to the experimental group or to the control group.

Some discrete variables are also referred to as categorical variables, because their values can be grouped into mutually exclusive categories. In the case of categorical variables, such as race, marital status, and religious orientation, the possible values of the variable include all of the possible categories of the variable. These categories are sometimes called **attributes.** For example, the attributes of the variable "marital status" could be defined as 1) single, never married, 2) married, 3) divorced, 4) widowed, not remarried, 5) living with a significant other, not married, and 6) other. In cases like this, it is always useful to include a category labeled "other" to include statuses that do not fit into the usual categories.

Measurement

It is important to clearly conceptualize and operationalize each variable in a research study. Variables must be defined in such a way that the researchers involved in the study, as well as those who utilize the research after it is published, understand the variables in the same way.

Likewise, the measurement of the variable must be defined in such a way that everyone involved in the study will measure it in exactly the same way each time it is measured. In addition, if another researcher wants to replicate your study at a later date, the measurement strategies should be clear enough so that they can be accurately replicated.

Measurement is a systematic process that involves assigning labels (usually numbers) to characteristics of people, objects, or events using explicit and consistent rules so that, ideally, the labels accurately represent the characteristic measured. Measurement is vulnerable to errors, both systematic and random. **Random measurement errors** are errors that occur randomly. **Systematic measurement error** is a pattern of error that occurs across many participants. We will cover random and systematic measurement errors in more detail in Chapter 3.

The goal of developing clear and accurate measurement procedures is to develop instruments with adequate reliability and validity. The **reliability** of a measurement is the degree of consistency of the measure. It reflects the amount of random error in the measurement. If a measure is applied repeatedly to the same person or situation, does it yield the same result each time? For example, suppose you use a bathroom scale to measure a person's weight. If it indicates the same weight each time the person steps onto it, then the scale or measure is reliable. It may not, however, be accurate. Suppose we go to the doctor's office and find out that the scale at home shows the person's weight as 20 pounds higher than his or her actual weight. The home scale is reliable, but not accurate based on the assumption that the doctor's scale is accurate.

The general definition of **measurement validity** is the degree to which accumulated evidence and theory support interpretations and uses of scores derived from a measure. The **validity** of a measurement refers to the accuracy of a measure. A measure can be reliable, as in the home bathroom scale above, but it may not be accurate or valid. Validity reflects the amount of systematic error in our measurement procedures. Suppose two observers of a student in a classroom are given a clear list of behaviors to count, for the purpose of measuring behaviors that correspond to symptoms of Attention Deficit Disorder (ADD). The researcher may have defined the behaviors that suggest the hyperactive symptoms of ADD, but failed to include the behaviors that suggest the inattentive symptoms of ADD. Therefore, the two observers would be able to consistently or reliably count the hyperactive behaviors of the study, but not the inattentive behaviors, and thus would not accurately or validly be assessing the total symptoms of ADD. Again, the key to creating measures that are both reliable and valid is to clearly conceptualize and operationalize each variable of interest in one's research study.

Levels of Measurement

The way a variable is operationalized will generally determine the level at which the data will be collected. If the variable "age" is defined by categories "0–10 years old," "11–20 years old," "21–30 years old," etc., then we would not know the actual age of the participants, but only these approximations. If participants are asked to enter their age, then we would know the actual number of years of age of each participant. If participants are asked to enter their birthdate, then we would know their age to the day. Determining what level of measurement we need for each variable is part of operationalizing a variable. Variables can be defined at four basic levels of measurement: nominal, ordinal, interval, and ratio. The level of measurement used for a variable determines the extent to which the value of a variable can be quantified.

Nominal. The first level of measurement is the **nominal** level. Nominal-level variables are categorical variables that have qualitative attributes only. The attributes or categories defined for a nominal variable must be exhaustive (meaning every response fits into one of the categories) and mutually exclusive (meaning each response fits into no more than one category). In other words, every possible response will fit into one and only one category defined for a variable. Let us return to the variable "marital status." Suppose the categories were defined as follows: 1 = single; 2 = married; 3 = divorced; and 4 = widowed. What if a person is divorced and now living as a single person? In this case, two categories could be selected, divorced and single; therefore, the categories are not mutually exclusive. What if a couple has been living together for 10 years and have 3 children together? Should they select "single?" In this case, there really are no categories that fit the couple's situation, thus the categories are not exhaustive. Other examples of nominal-level variables include race, gender, sexual orientation, and college major. As mentioned previously, a **dichotomous variable** is a variable measured at the nominal level that has only two possible attributes. Responses for dichotomous variables may include "yes/no," "true/false," "control group/experimental group," "male/female," and so on.

Ordinal. The second level of measurement is the **ordinal** level. Like the nominal-level variables, ordinal-level variables are also categorical, and the attributes must also be exhaustive and mutually exclusive. In addition to these characteristics, the attributes of an ordinal-level variable have an inherent order or ranking to them. For example, a variable "education level" could be defined to include the following attributes: 1 = less than high school education; 2 = graduated high school; 3 = some college, no degree; 4 = 2-year college degree; 5 = 4-year college degree; 6 = some graduate school, no degree; and 7 = graduate college degree. Unlike the earlier example of marital status, there is an inherent order or ranking to the attributes for this variable. If we listed them on a measurement instrument, it would always be listed in this order (or possibly in reverse order). In contrast, the attributes for the nominal-level variable, "marital status," could be listed in any order. Other examples of ordinal-level variables include "client satisfaction" (1 = extremely dissatisfied; 2 = dissatisfied; 3 = neutral; 4 = satisfied; 5 = extremely satisfied) and "level of agreement" (1 = completely disagree; 2 = somewhat disagree; 3 = neither disagree nor agree; 4 = somewhat agree; 5 = completely agree).

Interval. The third level of measurement is the **interval** level. While the first two levels are considered categorical variables, the values of an interval-level variable can be validly measured with numbers. Building on the requirements of the preceding levels, the attributes of interval-level variables are also exhaustive, mutually exclusive, and rank-ordered. In addition, the quantitative difference or distance between each of the attributes is equal. Looking at the variable, "education level," in the preceding example, there is not an equal amount of "education" between each of the categories. The difference between "graduated high school" and "some college, no degree" is not the same as the difference between a "4-year college degree" and a "graduate college degree." In an interval-level variable, there is equal distance between each attribute. For example, consider the scores on an IQ test that range from 50 to 150. The difference between a score of 50 and a score of 60 (10 points) is equal to the distance between a score of 110 and 120 (10 points).

Ratio. The fourth and final level of measurement is the **ratio** level. The attributes of a ratio-level variable are exhaustive, mutually exclusive, rank-ordered, and have equal distance between each attribute. One final

Table 2.1 Levels of Measurement

Level of Measurement	*Characteristics*	*Examples*
Nominal	Attributes are exhaustive Attributes are mutually exclusive	Race Gender Sexual orientation
Ordinal	Attributes are exhaustive Attributes are mutually exclusive Attributes have an inherent order	Client satisfaction Highest educational achievement Level of agreement
Interval	Attributes are exhaustive Attributes are mutually exclusive Attributes have an inherent order Differences between attributes are equal	IQ Score Temperature SAT Score
Ratio	Attributes are exhaustive Attributes are mutually exclusive Attributes have an inherent order Differences between attributes are equal Attributes have an absolute 0-point	Number of children Monthly income Number of times married

requirement yields a ratio-level variable: the presence of an absolute zero point. A variable can be measured at the ratio level only if there can be a complete absence of the variable. Examples include "number of children," "monthly mortgage payment," or "number of years served in prison." Note how all of these could be given the value of 0 to indicate an absence of the variable. In contrast, a temperature of 0 degrees Fahrenheit does not indicate an absence of temperature; therefore, temperature would be an interval-level variable rather than a ratio-level variable. See Table 2.1 for an overview of the levels of measurement described.

It is important to reiterate that a variable can often be defined at more than one level of measurement, depending on how it is conceptualized and operationalized. The researcher sometimes uses more than one variable at different levels of measurement in order to capture a concept more fully. See Table 2.2 for an example of how we can measure eating disordered behaviors using measurements at all four levels.

Statistical Levels of Measurement

The level at which a variable is measured determines what types of statistical methods can be used with that variable. Marlow (2011) points out

Table 2.2 Measuring a Concept at Different Levels

Variable/Item	*Attributes*	*Level of Measurement*
Have you purged during the past week?	Yes No	Nominal (dichotomous)
How often have you purged during the past week?	Not at all 1–2 times per day 3–5 times per day More than 5 times per day	Ordinal
Score on the *Eating Attitudes Test (EAT-26)* (Garner, Olmsted, Bohr, & Garfinkel, 1982)	Possible scores from 0 to 78	Interval
How many times have you purged during the past week?	Whole numbers 0 or greater	Ratio

that for the purposes of statistics, including working with computer statistical packages, interval- and ratio-level variables are generally combined into one category. For example, in SPSS, both interval- and ratio-level variables are referred to as scale-level variables. In addition, dichotomous variables (i.e., variables that have only two attributes or categories) are often treated separately from other nominal-level variables because there are statistical methods that are used with dichotomous variables that cannot be used when variables have more than two attributes. Thus, the four statistical levels of measurement are slightly different than the levels of measurement described previously; they include dichotomous, nominal, ordinal, and interval/ratio.

FREQUENCY AND PERCENTAGE DISTRIBUTIONS

Now that we have looked at the various steps in defining and measuring variables, we turn to three types of univariate statistical methods: 1) frequency and percentage distributions, 2) measures of central tendency, and 3) measures of variability. The most basic type of univariate statistical method is the tabular representation of **frequency** and **percentage distributions**. A frequency table displays a summary of the frequency of the individual values or range of values for a single variable. Table 2.3 shows a simple frequency distribution of the nominal variable “marital

Table 2.3 Simple Frequency Distribution

Marital Status	*Frequency of Responses*
Single, never married	59
Married	73
Divorced	42
Widowed	16
Living w/significant other	22
Other	5
Total	**217**

status." Clearly, it is much easier to summarize the data after seeing a frequency distribution than if we had to look at the values of all 217 individual cases.

Sometimes, it is helpful to group the individual values before placing them on a table. For example, suppose the variable "age" had been collected for 969 participants. Creating a simple frequency distribution to show ages might be cumbersome because there could be dozens of different ages represented in the sample. In this case, one can group the individual ages into categories before displaying them as a frequency distribution. See Table 2.4 for an example of a simple frequency distribution using grouped data.

In addition to frequencies, percentages are often included on a distribution table. In Table 2.5, we have added percentages to our previous example.

When displaying either ordinal-level variables or variables where data have been grouped before being placed in the distribution table, cumulative frequencies and percentages can sometimes be helpful. In a cumulative distribution table, the frequencies and/or percentages are accumulated in a separate column. In other words, the frequency (or

Table 2.4 Simple Frequency Distribution Using Grouped Data

Age	*Frequency of Responses*
Under 20 years	143
20 to 29 years	271
30 to 39 years	122
40 to 49 years	146
50 to 59 years	212
60 years and older	75
Total	**969**

Table 2.5 Percentage Frequency Distribution

Marital Status	*Frequency of Responses*	*Percentage of Responses*
Single, never married	59	27.2%
Married	73	33.6%
Divorced	42	19.4%
Widowed	16	7.4%
Living w/significant other	22	10.1%
Other	5	2.3%
Total	**217**	**100.0%**

percentage) in the first row is added to the frequency in the second row. This total is then added to the frequency in the third row, and so forth. In Table 2.6, we have added cumulative frequencies, percentages, and cumulative percentages to our previous example. When looking at this table, you can determine immediately how many participants are under age 30, under age 40, etc.

Before we move to the graphical representation of data, we offer one more suggestion. Tables should not be used to display data distributions that only involve two or three numbers. For example, if there are 34 males and 41 females in a sample, these data do not need to be displayed in a table. You can simply state the frequencies and percentages in the text of your manuscript. Some authorities suggest that six or fewer data points should be reported narratively, and not in a table or figure.

Graphical Representation of Distributions

Another way distributions of data can be represented is through a variety of graphs, such as line graphs, bar graphs, histograms, and pie charts.

Table 2.6 Cumulative Percentage Frequency using Grouped Data

Age	*Frequency of Responses*	*Cumulative Frequencies*	*Percentage of Responses*	*Cumulative Percentages*
Under 20 years	143	143	14.8%	14.8%
20 to 29 years	271	414	27.9%	42.7%
30 to 39 years	122	536	12.6%	55.3%
40 to 49 years	146	682	15.1%	70.4%
50 to 59 years	212	894	21.9%	92.3%
60 years and older	75	969	7.7%	100.0%
Total	**969**		**100.0%**	

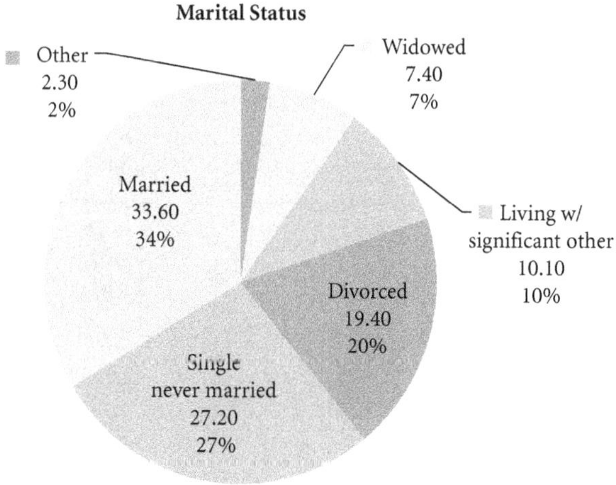

Figure 2.1 Pie Chart Showing Distribution from Table 2.5.

Probably the simplest of these is a **pie chart**. The percentage of the sample that has a certain value will be identified in the pie chart as a section of the pie representing that percentage. For example, look at the simple frequency distribution in Table 2.3. A pie chart showing these same data is displayed in Figure 2.1.

To create this pie chart using Microsoft Excel, you can follow these directions:

- Enter the labels for each value in Column 1.
- Enter the corresponding frequencies in Column 2.
- Select *Insert.*
- Select *Pie.*
- Select the format and color option that meets your needs. Avoid using subtle shades or colors, as these may be hard to read when reproduced on a black and white printer.

Like tables, distributions with only two or three data points generally should not be displayed in a graph. You do not need a pie chart to show that 45% of your sample is male and 55% is female. Simply stating these figures is sufficient.

Column and **bar graphs** are used to present numerical data in either horizontal or vertical columns. In some software, such as Excel, graphs

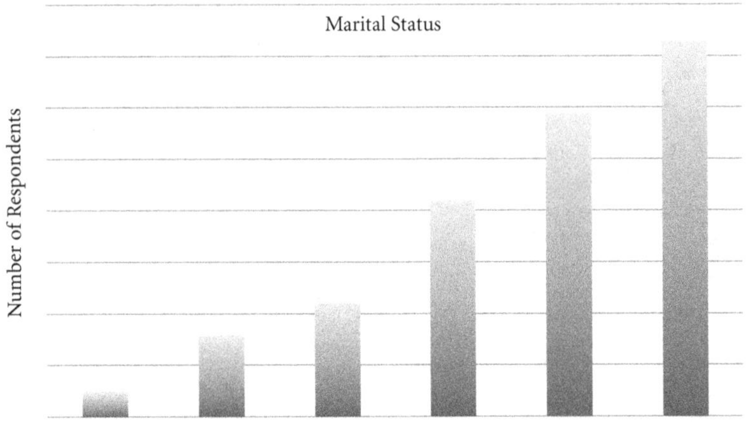

Figure 2.2 Column Graph Showing Distribution from Table 2.3.

that use horizontal bars to display the data are referred to as bar graphs, and graphs that use vertical columns are referred to as column graphs. In other software, the two terms are used interchangeably. Generally, the data represented in these types of graphs are discrete or categorical data, and the individual columns or bars are separated on the graphs. See Figure 2.2 for an example of a column graph and Figure 2.3 for a bar graph displaying the same marital status data as Figure 2.1.

If you are more interested in cumulative frequencies, percentages, or cumulative percentages than actual frequencies, these figures can also be

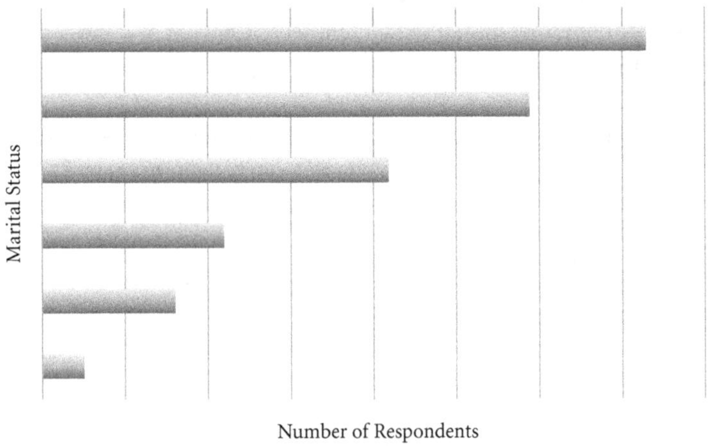

Figure 2.3 Bar Graph Showing Distribution from Table 2.3.

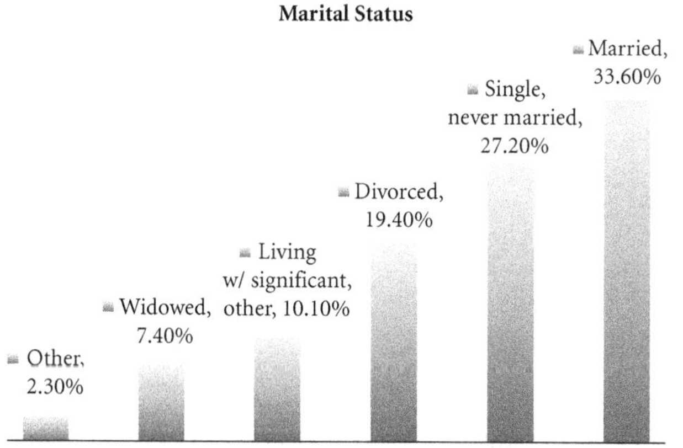

Figure 2.4 Column Graph Showing Percentages from the Distribution in Table 2.5.

displayed in a column graph. See Figure 2.4 for an example of using percentages in the graph. To create a bar or column chart in Excel, follow the provided on the previous page directions and select *Bar* or *Column* instead of *Pie,* then select the desired format.

A **histogram** is similar to a column chart, but is more often used to represent numeric data. The columns are generally placed adjacent on the chart with no space between values. A histogram of the data from Table 2.4 is displayed in Figure 2.5. Note that while these data are numeric (ages), the values have been grouped into categories (age ranges) and could also have been displayed using a bar or column graph. As in the other types of

Figure 2.5 Histogram Displaying Distribution from Table 2.4.

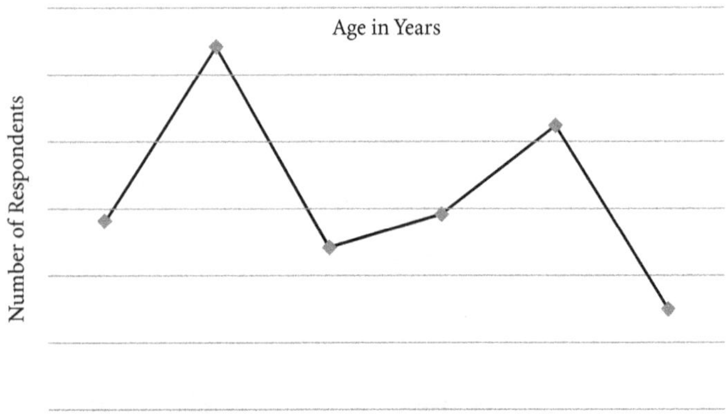

Figure 2.6 Line Graph Displaying Distribution from Table 2.5.

graphs, you can also display cumulative frequencies, percentages, or cumulative percentages rather than simple frequencies. How data are presented depends completely on the specific research study and the data being presented. There is no specific selection to create a histogram in Excel. You must select *Column* or *Bar*, then select the format in which the individual bars or columns are adjacent. Histograms are often used to check assumptions when conducting parametric tests or to ensure accuracy in model estimation. We will cover this in more detail in subsequent chapters.

One final type of graphical representation is the **line graph**. It is similar to the other types discussed previously except, instead of bars or columns points are placed on the graph and are connected by a line. Like histograms, line graphs are more often used to represent numeric data. See Figure 2.6 for an example line graph. To create a line graph in Excel, follow the previous directions and select *Line* instead of *Pie,* then select the desired format.

MEASURES OF CENTRAL TENDENCY

One of the most common ways we analyze and describe data is to compute **measures of central tendency**. Faulkner and Faulkner (2009) define a central tendency as "an estimation of the center of a distribution of values" (p. 157). The central tendency can also be viewed as the most common or typical value in a distribution. The most commonly used statistical measures of central tendency are the mode, the median, and the mean.

Mode

The **mode** is simply the value or attribute that occurs most frequently in a distribution of values. For example, suppose the test scores in your research class included the following scores:

41	60	68	68	72	75	77	77	80	80
82	85	88	88	88	90	90	91	94	96

The mode of this distribution would be 88, because this value appears three times and no other value appears more than twice. If two values or attributes appear the same number of times and more than any other value, then the distribution is bimodal. In the example above, if the 75 was a 77, then the values of 77 and 88 would both appear three times. In this case, the distribution would be considered **bimodal,** and 77 and 88 would be the modes. If three or more values appear an equal number of times and more than any other value, then the distribution is considered **multimodal.**

MEDIAN

Another measure of central tendency is the **median.** The median is the value in the middle of a distribution after all of the values have been ranked in order from smallest to largest. In other words, it is the value at which half of the data points are higher than the value and half are lower than the value. For example, look at the following SAT scores:

840	950	1020	1080	1110	1140	1150	1200	1250	1310	1440

There are a total of 11 values in this distribution. Therefore, the 6th value, or 1140 is the central value or median. There are five numbers less than 1140 and five numbers greater than 1140.

When the distribution includes an even number of values, there is one more step. Suppose we add one more value to the distribution above making it look like this:

840	950	1020	1080	1110	1140	1150	1200	1250	1310	1440	1460

Now there is no middle value. The scores of 1140 and 1150 are in the middle with five values less than 1140 and five values greater than 1150. In this case, you find the value that is halfway between the two center values. You do this by calculating the average or mean of these two values:

$$\frac{1140+1150}{2}=\frac{2290}{2}=1145$$

Therefore the median of this distribution is 1145.

Mean

The most commonly used statistical measure of central tendency is the **mean**, or statistical average. The mean of a population is signified by the symbol μ (mu), and is calculated by adding up all of the values in a population distribution and dividing by the number of values,

$$\mu = \frac{\Sigma X}{N}$$

where X represents the values in the population and N is the population size.

When calculating the mean of a sample (rather than a population), the formula is the same, using the sample size instead of the population size. The formula for calculating a sample mean is:

$$\overline{X} = \frac{\Sigma X}{n}$$

where $\overline{X}$ is the sample mean, X represents the values in the sample, and n is the sample size. Using the prior example of SAT scores, the sample mean would be calculated as follows:

$$\overline{X} = \frac{840+950+1020+1080+1110+1140+1150+1200+1250+1310+1440+1460}{12}$$

$$= \frac{13950}{12} = 1162.5$$

A variation of the mean is the **weighted mean**. In the weighted mean, instead of each value in the distribution contributing equally to the

Table 2.7 Calculating a Weighted Mean

Assignment	*Score*	*Percentage of Total*	*Weighting Factor*	*Total*
Homework #1	90	5%	.05	4.50
Homework #2	50	5%	.05	2.50
Homework #3	65	5%	.05	3.25
Homework #4	70	5%	.05	3.50
Attendance	100	20%	.20	20.00
Class Presentation	85	10%	.10	8.50
Midterm Exam	95	25%	.25	23.75
Final Exam	82	25%	.25	20.50
Final Grade				**86.50**

mean, some values contribute more than others, or are weighted more heavily. Teachers often use a weighted mean in order to calculate student grades. Consider a student's grades for a class as shown in Table 2.7. The mean of these grades would be calculated by summing the values (637) and dividing by the number of values (8) for a mean of 80. If we calculate the final grade this way, each of the individual scores is considered of equal importance. Suppose the teacher wants the grades weighted as shown in the second column of the table. In this case, each of the scores is multiplied by the weighting factor, and then these totals are summed to determine the student's final grade (or the weighted mean).

Selecting the Measure of Central Tendency

When should each of these measures of central tendency be used to describe a sample? In general, means are included in most research reports for all variables measured at the interval or ratio level. Means are generally preferred because every value in the distribution is included in its calculation. For example, many values in a distribution could change without affecting the mode or the median of the distribution. However, if just one value changes in a distribution, the mean will be affected.

Means are not, however, always appropriate. For example, it is mathematically inappropriate to calculate a mean for categorical variables measured at either the nominal or ordinal level. Modes are usually reported for categorical variables, or may simply be implied in a table or graph showing the frequency or percentage distribution of the data. In graphical representation, for example, the largest section of a pie chart would represent the mode or the value with the highest frequency within a distribution of values.

In cases where a data distribution includes one or more **outliers,** the mode or median is sometimes used in addition to or instead of the mean, because the mean might inaccurately represent the central or typical value of the distribution. An outlier is a value that is significantly greater or less than the other values in a distribution. For example, suppose the hourly wages of a group of clients are as follows:

$6.50	$7.00	$7.00	$7.10	$7.50	$8.00	$8.00	$8.50	$8.75	$9.00	$35.00

The mean of this distribution is calculated by summing the values ($112.35) and dividing by 11 (the total number of values), giving us a mean of $10.21. Because one of the clients earns an amount that is much larger than the other wages (i.e., $35.00), the mean is higher than all of the other 10 clients. Therefore, the mean does not accurately represent the typical or common hourly wage among this group of people. In this case, the mode (bimodal—modes are $7.00 and $8.00) or the median ($8.00) provide a more accurate representation of the central tendency of this distribution.

A **trimmed mean** is a variation of the mean that is sometimes used to reduce the effects of one or more outliers in the data. The trimmed mean is calculated by ignoring a small percentage (usually 5%) of the highest values and of the lowest values. The mean of the remaining values is then calculated using the standard formula.

Calculating Measures of Central Tendency using SPSS

In this demonstration, we explain how to calculate measures of central tendency (i.e., mean, median, and mode) in SPSS, version 19. We use a variable from the NELS data set that provides standardized reading scores. These data were collected during the first wave of the NELS study (1988). The name of the variable is BY2XRSTD. The name of the practice data set for this example is UNIVARIATE.SAV.

The following steps in SPSS are used to calculate the mean, median, and mode:

- Open the SPSS data file entitled, UNIVARIATE.SAV.
- Click on "Analyze" in the menu bar at the top of the data view screen.

- A drop down menu will appear. Move your cursor to "Descriptive Statistics."
- Another drop down menu will appear to the right of the first menu. Click on "Frequencies" to open the Frequencies dialog box.
- Highlight the variable you want to include and click on arrow to move it to the Variable(s) box. In this case, there is only one variable in the Frequencies dialog box, BY2XRSTD. Highlight it and move it to the Variable(s) box.
- Click on "Statistics" to the right of the Variable(s) box to open the "Frequencies: Statistics" dialog box.
- Check "mean," "median," and "mode," in the box entitled, "Central Tendency."
- Click "Continue" to close the dialog box.
- By default, a check mark appears in the box entitled, "Display frequency tables." Click on this box to remove the check mark.
- Click "OK." SPSS should now calculate the mean, the median, and the mode for the variable, BY2XRSTD.
- NOTE: A copy of this output is available on the companion website. The file is entitled, "Central tendency results.spv."

MEASURES OF VARIABILITY

In addition to the frequency and percentage distributions and the central tendencies of a distribution of values, it is also important to know how spread out or clustered together a group of data values are. **Measures of variability** (also known as **dispersion** or **spread**) are calculated to indicate the amount of variation that exists among the values of a distribution. We will look at three commonly used measures of variability: 1) range, 2) variance, and 3) standard deviation.

Range

The **range** of a distribution of values is simply the distance that encompasses all of the values of the distribution, including the minimum and maximum values. The range is calculated as follows:

$$\text{Range} = (\text{Maximum value} - \text{Minimum value}) + 1$$

The one is added in order to include both the minimum and maximum values in the range. For example, suppose we have the following group of test scores:

45	70	86	90	95	88	60	77	91	80	81

The range would be the maximum value (95) minus the minimum value (45) plus 1, or 51. This signifies that the data points in this distribution varied over a range of 51 points.

Interquartile range. The **interquartile range** is another statistic that can be used to minimize the effects of extreme outliers. The interquartile range is the range of values that falls between the 25th percentile and the 75th percentile. A **percentile** indicates the value of a distribution below which a certain percentage of the values fall. For example, if the 25th percentile of a distribution is 62, then 25% of the values are less than 62. If the 75th percentile of the same distribution is 85, then 75% of the values are less than 85. The interquartile range would be the 75th percentile (85) minus the 25th percentile (62) plus 1, or 24.

Variance

The **variance**, symbolized by σ^2, is a measure of variability that can be used with interval and ratio level variables only. It is calculated using the following steps:

- Subtract the mean of the distribution from each individual value or score,
- Square each of these differences,
- Sum the squares,
- Divide by the number of values if you are calculating the variance of a population or by the number of values minus one if you are calculating the variance of a sample.

The formula for variance looks like this:

$$\sigma^2 = \frac{\Sigma(X - \mu)^2}{N}$$

where X represents all of the values in the population, μ is the mean of the population, and N is the size of the population. To calculate the variance for a sample, the sample mean replaces the population mean, and the population size (N) is replaced by the sample size minus one ($n - 1$). The sample variance formula looks like this:

$$s^2 = \frac{\Sigma(X - \overline{X})^2}{n - 1}$$

We return to the test scores from the previous example:

45	70	86	90	95	88	60	77	91	80	81

In order to calculate the variance, you first must calculate the mean of this distribution.

$$\overline{X} = \frac{\Sigma X}{n} = \frac{863}{11} = 78$$

The next step is to calculate the sum of the squares (the numerator of the formula for variance). See Table 2.8 to see how the sum of the squares is calculated. The calculation for the variance is then simply:

$$s^2 = \frac{2257}{10} = 225.7$$

Table 2.8 Calculating Variance

Test Scores	*Mean*	*Deviation from the Mean*	*Square*
45	78	−33	1,089
70	78	−8	64
86	78	8	64
90	78	12	144
95	78	17	289
88	78	10	100
60	78	−18	324
77	78	−1	1
91	78	13	169
80	78	2	4
81	78	3	9
Sum of the Squares			**2,257**

Standard Deviation

The final measure of variability we will consider is the **standard deviation**. The standard deviation indicates the average distance from the mean that the individual values in a distribution fall. If the values in a distribution cluster tightly around the mean, then the standard deviation will be small. If the values vary widely from the mean, then the standard deviation will be large. Like the variance, the standard deviation requires interval- or ratio-level variables. Standard deviation is probably the most commonly used measure of variability and is a key component in many types of statistical analyses.

To calculate the standard deviation, you simply take the square root of the variance. Continuing with the previous example, the standard deviation would be calculated as follows:

$$s=\sqrt{s^2}=\sqrt{225.7}=15.02$$

In other words, the values in this distribution differ from the mean by an average of 15.02 points. This figure is quite high because the test scores used for this example had a wide range. If the scores had been clustered more closely around the mean, then the standard deviation would have been smaller.

Calculating Measures of Variability using SPSS

In this demonstration we explain how to calculate measures of variability (i.e., standard deviation, variance, and range [including the minimum and maximum values]) in SPSS, version 19. We must use a numeric variable in order to calculate these statistics. In this case we use a variable from the NELS data set that provides standardized reading scores. These data were collected during the first wave of the NELS study (1988). The name of the variable is BY2XRSTD. The name of the practice data set for this example is UNIVARIATE.SAV.

The following steps are used to calculate standard deviation, variance, and range:

- Open the SPSS data file entitled, UNIVARIATE.SAV.
- Click on "Analyze" in the menu bar at the top of the data view screen.

- A drop down menu will appear. Move your cursor to "Descriptive Statistics."
- Another drop down menu will appear to the right of the first menu. Click on "Frequencies" to open the Frequencies dialog box.
- Highlight the variable you want to include and click on the arrow to move it to the Variable(s) box. In this case, there is only one variable in the Frequencies dialog box, BY2XRSTD. Highlight it and move it to the Variable(s) box.
- Click on "Statistics" to the right of the Variable(s) box to open the "Frequencies: Statistics" dialog box.
- Check "standard deviation," "variance," "range," "minimum," and "maximum" in the box entitled, "Dispersion."
- Click "Continue" to close the dialog box.
- By default, a check mark appears in the box entitled, "Display frequency tables." Click on this box to remove the check mark.
- Click "OK." SPSS should now calculate the standard deviation, the variance, the range, the minimum score, and the maximum score for the variable, BY2XRSTD.
- NOTE: A copy of this output is available on the companion website. The file is entitled, "Dispersion results.spv."

3

Inferential Statistics

INTRODUCTION

In the previous chapter, we looked at descriptive statistical methods used to summarize and describe our observations. This chapter focuses on **inferential statistics**. These are statistical methods that use data we have observed in a sample to make hypotheses or predictions about data that have not been observed directly in the larger population. As indicated in Chapter 2, we calculate statistics from our sample as estimates of population parameters. Inferential statistics make it possible for us to predict whether relationships between variables found in our sample will hold true for the larger population. In this chapter, we will first discuss different types of hypotheses, and how different relationships between variables are used to develop these hypotheses. Then we will examine how data from a sample can be used to accurately predict information about the larger population.

RESEARCH AND NULL HYPOTHESES

Inferential statistics are generally used to determine if there is a relationship between two variables. For example, are boys more likely to return to the juvenile justice system after their first offense than girls? We want to know if gender and recidivism are related. Or was the depression level among the participants who completed the cognitive therapy sessions lower than the participants who used medication to treat their depression? We are trying to determine if there is a relationship between depression level and type of treatment.

Null Hypotheses

In research, the convention is to test the *absence* of a relationship between the variables. We try to determine if the difference between the values could be explained by sampling error. The hypothesis that presents the assumption that there is no relationship between the variables is called the **null hypothesis**. The null hypotheses that would coincide with the examples above would be:

> *There is no relationship between gender and recidivism among first-time juvenile offenders.*
>
> *There is no difference between the depression levels of the participants who completed the cognitive therapy sessions and those who used medication to treat their depression.*

After conducting our inferential statistical tests, we either reject or fail to reject the null hypothesis. If the null hypothesis is rejected, the alternative hypothesis that predicts a relationship between the variables may be supported. This alternative hypothesis is generally referred to as the research hypothesis.

Research Hypotheses

The **research hypothesis** states the assumption that there is a relationship between the variables of interest. The research hypotheses that might be used along with the null hypotheses above would be:

> *Male first-time juvenile offenders will be more likely to recidivate than female first-time juvenile offenders.*
>
> *There will be a difference between the depression levels of the participants who completed the cognitive therapy sessions and those who used medication to treat their depression.*

Do you notice a difference between these two research hypotheses? The first example predicts not only a relationship between the two variables, gender and recidivism, but also predicts the direction of the relationship. In other words, it specifically predicts that the boys will be *more likely* to recidivate than the girls. This type of hypothesis is called a **directional** or **one-tailed hypothesis.** The second example predicts that there will be a relationship between type of treatment and depression level, but it does not predict which group of participants will have lower depression levels and which group will have higher levels. In other words, it does not predict the direction of the relationship. Therefore, this type of hypothesis is called a **nondirectional** or **two-tailed hypothesis.** The second example hypothesis could be restated as the following directional hypothesis:

> *The participants who completed the cognitive therapy sessions will have lower depression levels than those who used medication to treat their depression.*

Some researchers consider a hypothesis to be *stronger* if it predicts the direction of the relationship between the variables. The advantage of finding evidence to support a one-tailed hypothesis test is that it is more powerful than a two-tailed hypothesis test because the results are in the predicted direction. Others argue that a nondirectional hypothesis should be used because it leads to a more conservative statistical test. The advantage of a two-tailed hypothesis test is that it allows detection of an unanticipated result. We discuss the significance of the one-tailed and two-tailed hypothesis, and why the two-tailed test is more conservative, later in this chapter. Other characteristics, such as falsifiability, have also been discussed with relation to the strength or weakness of research hypotheses.

In addition to predicting a relationship between variables and sometimes the direction of this relationship, a research hypothesis also predicts the nature of the relationship between the variables. The three main types of relationships between variables are association, correlation, and causal relationships. A relationship of **association** is predicted between variables that are measured at the nominal or ordinal levels. Association simply predicts that certain value categories of one variable will be associated or found more often with certain value categories of another variable. For example:

> *Girls are more likely than boys to pass the competency exam to enter middle school.*

The first variable, *gender*, includes two value categories, *boy* and *girl.* The second variable, the outcome of the competency exam, is also dichotomous with two value categories, *pass* and *fail.* This hypothesis predicts that the value "girl" will be found with the value "pass" more often than the value "boy" will be found with the value "pass."

Correlational relationships can be either positive or negative, and are found between variables that are measured at the interval or ratio levels. A **positive correlation** between variables predicts that high values of the first variable will be found with high values of the second variable, and low values of the first variable will be found with low values of the second variable. A **negative correlation** between variables predicts that high values of the first variable will be found with low values of the second variable, and low values of the first variable will be found with high values of the second variable. An example hypothesis predicting a positive correlation is:

Students with higher SAT scores will earn higher GPAs during their freshman year in college.

An example hypothesis predicting a negative correlation is:

Students with higher absentee rates from university classes will have lower GPAs.

Association and correlation *do not* necessarily predict a causal relationship between the two variables. In other words, they are not inferring that the values of one variable cause the values of the second variable. They only predict that certain values or value categories of the first variable will be found with certain values or value categories of the second variable. Hypotheses that infer a **causal** relationship predict that values of one variable actually cause or directly influence the values in another variable. For example:

Clients who participate in the self-esteem workshop will score higher on the self-esteem inventory than those clients who do not participate in the workshop.

Here the researcher predicts that participation in the self-esteem workshop will influence the client's score on the self-esteem inventory. To establish causality (e.g., the self-esteem intervention causes increased self-esteem), the criteria described in the Introduction (Chapter 1) must be met. Next we will look at the different types of variables used in hypotheses and the roles played by each type.

Types of Relationships between Variables

In Chapter 2, we described how to define variables and the different levels at which variables can be measured. In this section, we will look at the different ways variables may be related to each other. The most common terms used in research to describe the relationship between variables are independent and dependent variables. An **independent variable (IV)** is a variable that is predicted to explain or cause the values of other variables in the research study. The **dependent variable (DV)** is the variable that we believe will be influenced by the IV. For example, if half of the participants in a research study receive cognitive therapy to treat their anorexia nervosa and the other half are placed in a comparison group and receive no treatment, whether or not the participants received the cognitive therapy is the IV, and some measure of anorexia-nervosa symptoms is the DV.

Variables are not defined as either independent or dependent based on the concepts they represent. Rather, they are either independent or dependent based on the purpose and context of the research study in which they are defined. For example, in the above example, a measure of the severity of anorexia-nervosa symptoms was defined as the DV. Another study may look at the severity of anorexia-nervosa symptoms of a group of clients, and try to determine whether more severe symptoms of anorexia nervosa lead to higher levels of depression. In this example, the severity of anorexia-nervosa symptoms is defined as the IV and depression level as the DV.

When the relationship between variables being predicted is an association or correlation (not causal), the terms predictor and outcome variable may be used instead of dependent and independent. The **predictor variable** is hypothesized to predict the values of the **outcome variable**. If we predict an association or correlation, but are *not* hypothesizing that one variable can be used to predict the other, these terms are not used.

Other variables are considered **moderator** or **mediator variables.** A **moderator variable** is a variable that influences the strength of the relationship between an IV and a DV, while a **mediator variable** (i.e., **intervening variable**) actually explains or accounts for all or part of the relationship between the IV and DV. For example, suppose we find a relationship between socioeconomic level and prenatal care received by single mothers. Age would be a moderator variable if the relationship between socioeconomic level and prenatal care is stronger for younger women and weaker for older women. Amount of social support would be a mediator variable if it explains all or part of the relationship between socioeconomic level and prenatal care.

If the researcher attempts to determine the effect the IV has on the DV, it is important to consider how other variables may affect the DV. For example, suppose we are evaluating a program that is attempting to change peoples' attitudes toward immigrants in a small community. We might predict that the program may be more effective with people who have a higher educational attainment. In this case, education level would be considered a possible moderating variable. When the researcher recognizes a possible moderating variable in a research study, and attempts to control its effects on the DV, it is referred to as a **control variable**. The researcher controls the effects of these variables by holding their effects constant while observing the IV's influence on the DV.

Variables that have a more specific type of effect on the relationship between the IV and DV are suppressor variables. A **suppressor variable**, also referred to as an **obscuring variable,** is a type of control variable. The way it works is that the relationship between the IV and the DV is only statistically significant when the suppressor variable is included in the analysis. For example, in their study on medical rehabilitation outcomes among older adults, MacNeill, Lichtenberg, and LaBuda (2000) discovered that the IVs, age and education, were statistically related to the DV, return to independent living, only when the analyses included the variables that measured cognitive ability.

A c**onfounding variable** is an extraneous variable (i.e., not included in the model) that correlates with both the independent and dependent variables and can lead to incorrect conclusions about the nature of the relationship between the IV and DV. An example of a possible confounding variable in the MacNeill et al. (2000) study is family support. This variable is likely to have an impact on the DV—return to independent living. If so, and if the authors do not include the family support variable in the analysis, then the variable is referred to as a confounding variable.

SHAPES OF DATA DISTRIBUTIONS

In the previous chapter, we showed various ways data can be represented visually, such as line graphs, pie charts, and bar graphs. These graphs are used to show the *shape* of the distribution of the values of a variable. There are a multitude of different shapes that a distribution may follow. For example, some distributions are bimodal, meaning there are two values that occur most frequently. A bimodal distribution would have two peaks rather than just one. A distribution can also look like a wave and be multimodal.

Some distributions are fairly flat such that the spread of values is widely scattered across the curve, with almost equal frequencies for each value of the variable. Other distributions may be very peaked with one or a very few values representing the majority of the cases. The **kurtosis** is the degree to which a distribution is flat, peaked, or bell-shaped. A distribution that has a rounder, wider peak around the mean is referred to as platykurtic. The values are less concentrated around the mean than in a normal distribution. A distribution that has a sharper, higher peak around the mean is referred to as leptokurtic. The values are more concentrated around the mean than in a normal distribution. A distribution that approximates the shape of a bell curve is referred to as mesokurtic. The most well-known example of a mesokurtic distribution is the normal distribution. See Figure 3.1 for an illustration of each of these distributions.

The **skewness** of a distribution refers to the degree of asymmetry of a distribution. A positively skewed distribution refers to a distribution in which the majority of observations fall at the lower end of the range of values with a longer tail on the right. A negatively skewed distribution refers to a distribution in which the majority of observations fall at the higher end of the range of values with a longer tail on the left. As an example, suppose students take a pretest on course material before they are exposed to the material in class. These scores would likely tend to be positively skewed, with the majority of the students scoring poorly on the test, with a few students scoring in the moderate range, and a few scoring

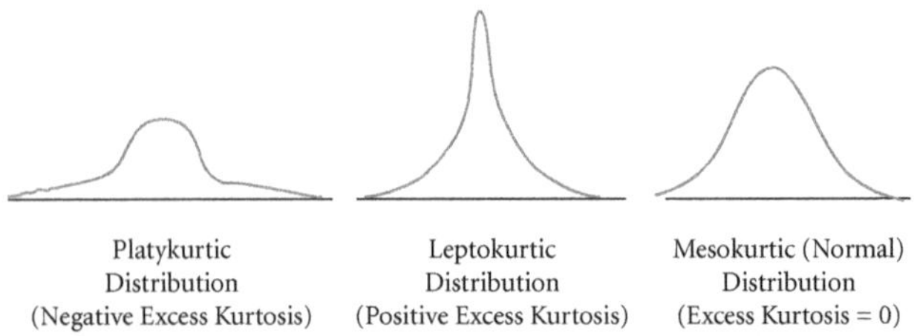

Figure 3.1 Example of Differing Kurtosis.

very high (see Figure 3.2). After the material is presented and the students take a posttest, the scores will likely create a negatively skewed distribution, with the majority of students scoring well, then gradually tapering off with a few students scoring poorly (see Figure 3.3). As with kurtosis, we can also calculate a statistic that indicates the degree of skewness of a distribution. Later on we will discuss the implications of the shapes of distributions in statistical estimation.

The Normal Distribution

The most common distribution of data is called the **normal distribution.** We are going to concentrate our discussion on the normal distribution or curve, because it is involved in many aspects of statistical analyses of data. Recall from the Introduction (Chapter 1) that one assumption in parametric statistical testing is that the data are normally distributed.

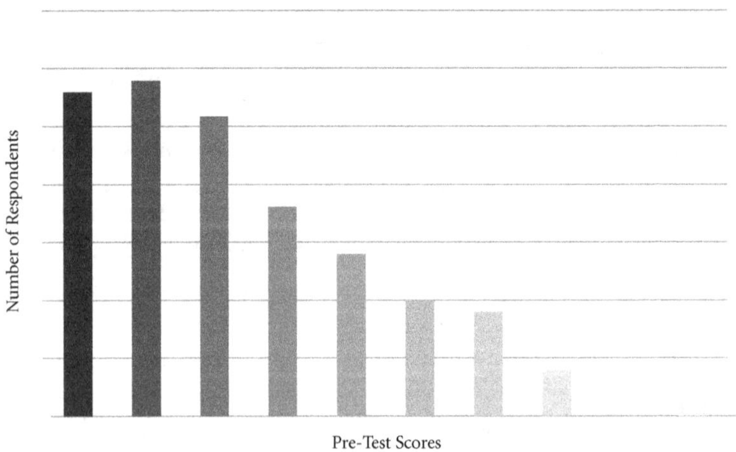

Figure 3.2 Example of Positively Skewed Distribution—Pretest Scores.

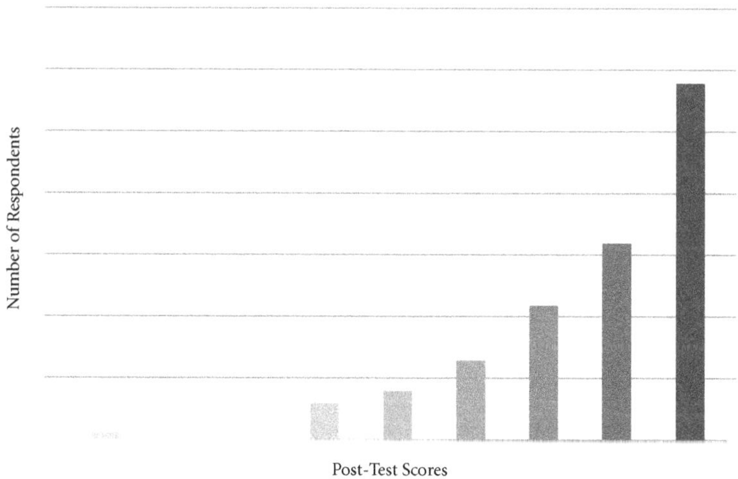

Figure 3.3 Example of Negatively Skewed Distribution—Posttest Scores.

You will learn about this and other ways in which the normal distribution is critical in inferential statistics in subsequent chapters. A normal distribution is relatively bell-shaped and symmetrical. Most of the distribution's data points cluster near the center. Theoretically, the ends of a normal curve continue toward infinity in both directions, getting closer and closer to the *x*-axis but never actually touching it in a perfect normal distribution. The mean, median, and mode of the distribution are at the center or peak of the distribution in such a way that they are all equal (see Figure 3.4).

Most variables that occur in nature, such as the height and weight of animals or the size of fruits and vegetables, follow a normal distribution. Other variables that are not "natural," such as the gas mileage of all American-made vehicles, test scores of a social work class, or depression levels of university students, will generally not follow a normal distribution.

The mean and median of the distribution are at the center, which is also the mode or highest point of the normal distribution. The shape of the distribution depends on the standard deviation of the data. If the standard deviation is high, the data will be spread out into a wider, flatter curve. If the standard deviation is low, the shape of the distribution will be more peaked and narrow (see Figure 3.5). This is similar to the previous discussion about platykurtic and leptokurtic distributions. All

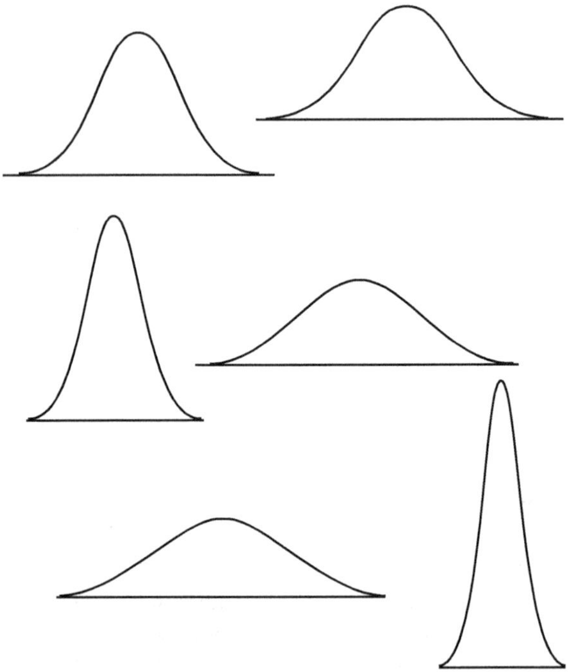

Figure 3.4 Examples of Normal Distributions.

normal distributions have the same degree of kurtosis. A normal distribution with a lower standard deviation will have a higher peak in the middle, but this does not make it a leptokurtic distribution. A normal curve is said to have an excess kurtosis of 0. The higher the positive kurtosis (above 0) a distribution has, the more leptokurtic the distribution is. Conversely, the more negative kurtosis (below 0), the more platykurtic the distribution is. A normal curve retains the bell-shaped curve, which is characteristic of all normal distributions.

An even more interesting and useful characteristic of the normal distribution is that the horizontal axis of a normal distribution can be divided into six units based on the distribution's standard deviation. This is possible because the distribution is symmetric around the center line (mean/mode/median) of the curve. Look at Figure 3.6 for the proportions of the normal curve. Virtually all (99.74%) of the values in the normal distribution fall within these six units, with 49.87% falling between the mean and plus three standard deviations,

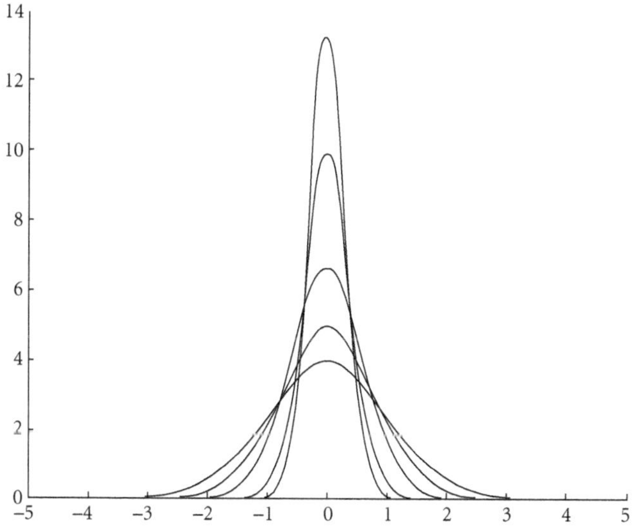

Figure 3.5 Normal Distributions with Equal Means and Different Standard Deviations.

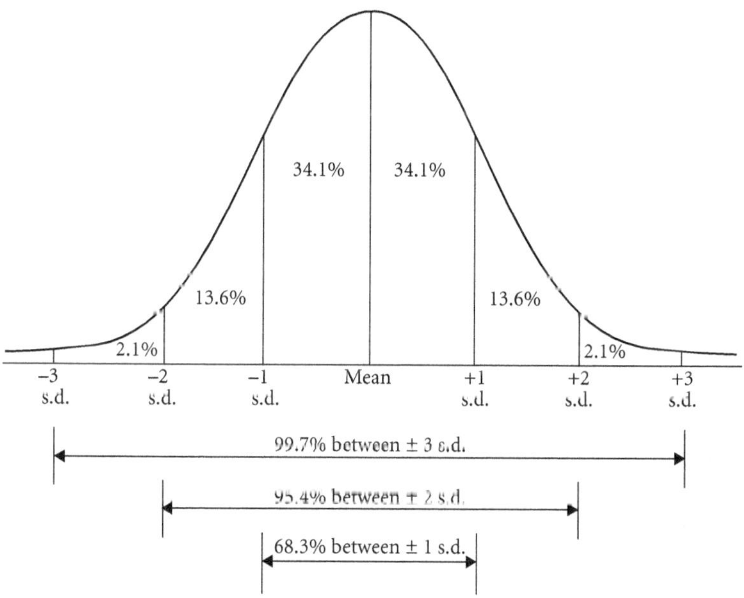

Figure 3.6 Proportions of the Normal Distribution.

and 49.87% falling between the mean and minus three standard deviations.

As you can see from the figure of a normal distribution, the percentage of values that fall within each of the six units is known. This gives us a significant amount of information about data that follow a normal distribution. For example, if the mean of a normal distribution is 12 and the standard deviation is 2, we can deduce that 34.13% of the values will fall between the mean (12) and plus one standard deviation from the mean (12 + 2 = 14). In addition, 34.13% of the values fall between the mean (12) and minus one standard deviation from the mean (12 – 2 = 10). We can similarly calculate the percentage of values between the mean and plus two standard deviations (47.72%) and the mean and plus three standard deviations (49.87%). Therefore, 68.26% of the values fall between plus (12 + 2 = 14) and minus (12 – 2 = 10) one standard deviation from the mean, 95.44% of the values fall between plus (12 + 4 = 16) and minus (12 – 4 = 8) two standard deviations from the mean, and 99.74% of the values fall between plus (12 + 6 = 18) and minus (12 – 6 = 6) three standard deviations from the mean.

Percentiles and *z* scores. We can also use this figure to estimate the percentile of a value in a normal distribution. Remember, the percentile of a value is the percentage of all values in a distribution that are less than the value. Using the same mean (12) and standard deviation (2), we can calculate the percentile of the value 16. First, determine the percentage of values that lies between the mean and 16. Because 16 is plus two standard deviations from the mean, 47.72% of the values lie between 12 and 16. We must also add in the 50% of the values that lie to the left of the center line or mean. Therefore the value of 16 is at the 97.72nd (47.72 + 50) percentile in this distribution (see Figure 3.7). To calculate the percentile of 6, we first determine that 6 is minus three standard deviations from the mean. Therefore, 49.87% of the values lie between 6 and 12. To determine the percentage of values that lie to the left of 6, we simply subtract 49.87% from all of the values to the left of the mean, which is 50%, and get .13%. Therefore, 6 is at the .13th (50–49.87) percentile (see Figure 3.8).

We can only use Figure 3.6 to calculate the exact percentile of a value if it is an even multiple of the standard deviation from the mean. In all other cases, the raw value or score must be converted to what is referred to as a ***z* score** or **standard score;** the *z* score is then used to determine the

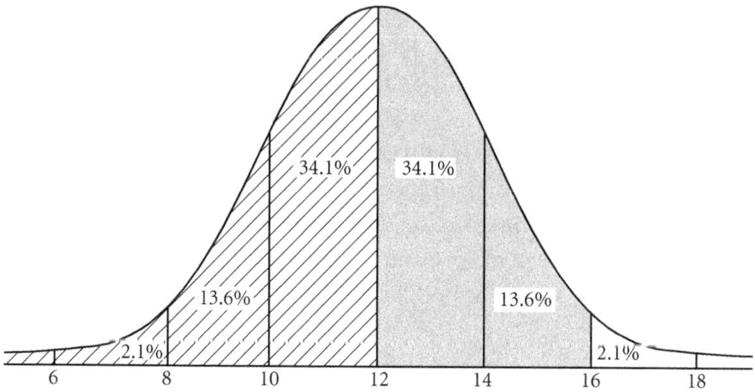

Figure 3.7 Calculating the Percentile of a Value at Plus Two Standard Deviations from the Mean.

percentile. The *z* score is simply the number of standard deviations a raw score falls above or below the mean. If the raw score falls below the mean, the *z* score will be negative, and if the raw score falls above the mean, the *z* score will be positive. The following formula is used to calculate the *z* score:

$$z \text{ score} = \frac{\text{raw score - mean}}{\text{standard deviation}}$$

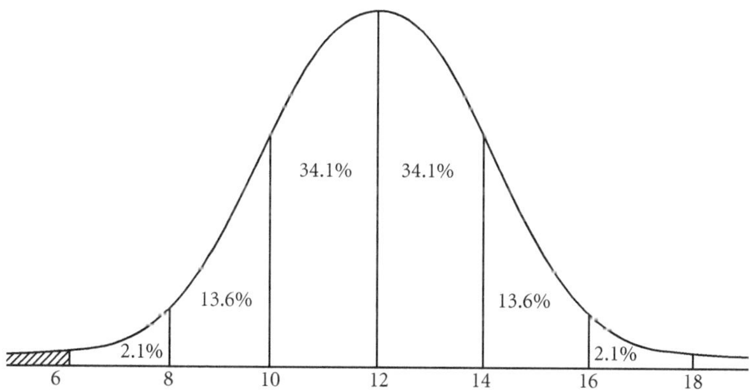

Figure 3.8 Calculating the Percentile of a Value at Minus Three Standard Deviations from the Mean.

In addition to calculating the percentile of a raw score, *z* scores allow us to compare raw scores from different samples. For example, suppose you received a 74% on your research mid-term and an 84% on your statistics mid-term. At first glance, you would assume that you performed better on the statistics test. However, suppose the scores on the research mid-term were normally distributed with a mean of 71% and a standard deviation of 6%, and the scores on the statistics mid-term were normally distributed with a mean of 86% and a standard deviation of 2%. The calculations for the *z* scores for both of these tests would look like this:

$$z \text{ score (research mid-term)} = \frac{74-71}{6} = \frac{3}{6} = 0.5$$

$$z \text{ score (statistics mid-term)} = \frac{84-86}{2} = \frac{-2}{2} = -1.0$$

In other words, you scored 0.5 standard deviations above the class mean on the research mid-term and 1.0 standard deviation below the class mean on the statistics mid-term. Therefore, you could argue that you performed better on the research exam, relative to the other students in the class.

When a *z* score is a whole number, we can return to Figure 3.6 to calculate the percentile for the raw score. In the above example, the *z* score for the statistics exam was −1.0. Looking at Figure 3.6, we can easily determine that 34.13% of the values fall between the raw score (84%) and the mean (86%). To calculate the percentile, we simply subtract 34.13 from 50 to determine that 15.87% of the class scored less than 84% on this test.

How do we calculate a percentile for a *z* score that is not a whole number? To convert fractional *z* scores to percentiles, we must turn to a table that displays the areas under the normal curve (go to the website http://www.statsoft.com/textbook/distribution-tables/ for an example standard normal *z* table). First ignoring the sign of your *z* score, find the first digits of the *z* score in the left-hand column. Next, move across the row to find the second decimal. The value that appears in the table is the area between your raw score and the mean. You then have to convert this value to the percentile.

First, multiply the area from the table by 100 to give you the percentage of values that falls between your raw score and the mean. If your *z* score is positive (i.e., the raw score is greater than the mean), the score on the table represents the percentage of values from the mean to the right until you reach the raw score. Therefore, you add 50 to the percentage from the chart to account for the percentage of values that lie to the left of the mean or center line. If your *z* score is negative (i.e., the raw score is less than the mean), the table represents the percentage of values from the mean to the left until you reach the raw score. Therefore, to calculate the percentile (the percentage of values that falls below your raw score), you subtract the percentage identified in the table from 50.

Returning to our prior example, how do we calculate the exact percentile for our raw score of 74 and a *z* score of 0.5? Looking at the standard normal *z* table, look down the first column until you get to 0.5. Go to the right to the next column which represents the *z* score, 0.50. The value in the table is .1915. Multiply this value by 100 to get 19.15. Because the *z* score is positive, we add 50 to 19.15 to get a percentile of 69.15. Therefore, we scored at the 15.87th percentile on our statistics mid-term and at the 69.15th percentile on the research mid-term, making another argument that we performed better on the research test even though our raw score was higher on the statistics test. This example highlights one way *z* scores are used to determine percentile ranks when dealing with normal distributions. In general, standard scores follow the same distributions as the variables on which they are based, and can be used in a variety of more advanced statistical techniques that are not covered in this text.

The Assumption of Normality in Statistical Estimation

In subsequent chapters, we describe procedures for statistical estimation using multiple linear regression, Analysis of Variance and Covariance, and path analysis. Each method is a parametric test that assumes a normal distribution of the dependent variable. We can determine the extent to which a distribution is normal by plotting the observations on a histogram (discussed in Chapter 2) or calculating skewness and kurtosis statistics in SPSS. In a normal distribution, skewness and kurtosis statistics equal 0. The further these values are from 0, the more likely it is that the distribution is not normal (i.e., skewed and/or kurtotic). By convention, skewness statistics with values absolute values greater than 1.96 deviate beyond acceptable limits of normality. Kurtosis statistics with values absolute values greater than 3.29 deviate beyond acceptable limits of normality.

RESEARCH SAMPLING

If we only needed statistics to summarize the data we have collected and to describe the distribution of the values of one or more variables within our sample or population, we could use the information in the preceding sections of this book to meet our needs. However, we often want to discover if there is a relationship between variables or between different samples. For example, we may want to know if a drug abuse prevention program is more effective when we offered it to eighth grade students or sixth grade students. Or is a self-esteem intervention delivered to middle school youth more effective if given in a co-ed class or in classrooms where boys and girls are separated? Or, does a person's age affect his or her ability to learn new job skills at a job training center? This is where we turn to inferential statistics.

There are really two steps involved in answering these questions. The first step is to determine whether there is a relationship between the variables within the sample or samples we actually have studied. We will return to this question later in the chapter, and we will cover statistical techniques involved in answering this question in the remaining chapters of this book. The second step is to assess a level of confidence that the findings based on our sample would hold true with other samples drawn from the same population. This is an indicator of **external validity** of a study—the extent to which results can be generalized to other samples. For instance, if we discover that the self-esteem program we offered in our middle school was more effective when we divided the girls and boys than when we placed them in a group together, how confident are we that this would hold true for another group of middle school students in the middle school, or to other middle school students arounds the state, or the country? These are much larger questions, and relate to the next topic we will discuss: research sampling.

Research sampling is the process of obtaining participants for our research study. The sample may consist of people, couples, families, communities, schools, states, organizations, or almost anything else researchers might be interested in studying. In fields other than social work, a sample may consist of rats, cells, germs, or rocks. Researchers select a sample, because it is generally impractical to study an entire population. However, it is not impossible. If a historian wants to analyze the leadership styles of the United States presidents, he may decide to include all of the presidents. However, if the same historian wants to analyze the leadership styles of all of the men and women who have served in the United States Congress, he would most likely select a sample. It is less expensive and takes much less time to study one or more samples from a population rather than the entire population.

How are the areas under the curves for a normal distribution related to the question of how confidently we can generalize our findings? We stated earlier that inferential statistics make it possible for us to determine if findings from our sample can be generalized to the larger population. The accuracy of these generalizations depends on how closely our sample represents the population. If the sample accurately represents the population from which it is drawn, then generalizations made based on the sample will be accurate as well, and the study is thought to have strong **external validity**. The representativeness of the sample depends primarily on whether the sample is randomly selected from the larger population.

Random sampling refers to sampling techniques in which each element in the population has a known and equal chance of being selected. Random-sampling procedures are used to strengthen a study's external validity, or the extent to which the results can be generalized to other samples outside of the study. **Nonrandom sampling** refers to sampling techniques in which each element in the population has an unknown and usually unequal chance of being selected (for a more detailed discussion of various random and nonrandom sampling techniques, see Krysik & Finn, 2010; Rubin & Babbie, 2008; or Yegidis, Weinbach, & Myers, 2012).

Before we move on, it is useful to clarify the difference between random sampling and random assignment. Random sampling involves the process of selecting a sample of study participants from the larger population. Random sampling is used to strengthen external validity, or the extent to which the findings from the sample can be generalized to the larger population. On the other hand, **random assignment** occurs after the study's sample has been selected, and involves assigning participants to experimental or control groups on the basis of chance. Random assignment is used to strengthen a study's **internal validity**, or the extent to which explanations for the study's findings (i.e., effect) can be attributed to the experiment (i.e., cause), and alternative explanations for the cause and effect relationship can be ruled out.

If our sample is selected from the population using a nonrandom sampling method, there is likely to be a difference between our sample and the population from which it was drawn. This difference, called the **sampling error,** may be substantial. If the sample is selected using a random-sampling method, the sampling error will likely be much smaller. For example, suppose a university professor wanted to survey the students in her university. If she surveyed only students in her social work classes, the sample would probably differ substantially from the larger university. If she surveyed students in every department on campus, but selected only those students that were most convenient to survey, the sample would more closely represent the university students, but would still most likely differ from the population of all university students. If, on the other hand, the professor got a list of all the university's students and randomly selected her sample from the list, the sample would more closely represent the population of university students.

If random sampling is used, does it guarantee that the sample will perfectly represent the population? Unfortunately, it does not. We can assume that even a sample drawn randomly will differ in some aspects from the population from which it was drawn. We may find that a relationship between variables exists in the sample, when, in fact, the same relationship does not exist in the population. On the contrary, a relationship may not exist in the sample, while it does exist in the population. Or, the relationship may exist in both the sample and the population or may not exist in the sample or the population.

There are many reasons why a relationship found between variables in a sample may not reflect the true relationship in the population. Three common reasons why researchers misjudge relationships between variables are measurement error, sampling bias, and sampling error.

Measurement error. If we measure the variables of interest within our sample incorrectly or in such a way as to introduce bias, then our conclusions are likely to not hold true for the larger population. They may not even hold true for our sample! There are two basic types of measurement error: systematic-measurement error and random-measurement error. **Systematic-measurement error** represents discrepancies between measured and actual ("true") values of a variable, where the values of the measured variable are more likely to be higher or lower than the actual values of the variable. Systematic-measurement errors are caused by flaws in the measurement process, and they lead to over- or underestimates of the actual values of a variable. This type of error is also known as "bias" in measurement. For example, suppose the researcher wishes to compare outcomes of an ADHD intervention with a group of second grade children to the same intervention with a group of fourth grade children. She creates a behavioral observation protocol to count the number of times the children disrupt the class. The researcher decides to observe each class immediately after their recess periods in order to make the observation periods as similar as possible. However, the second grade children have their recess at 9 AM while the fourth grade children have their recess at 2 PM. The researcher concludes that the second grade children responded much better to the intervention. In reality, the ADHD medications may simply be wearing off for many of the fourth grade children by 2 PM making them appear to react much more poorly to the intervention.

Random-measurement error is just that: random. It does not follow a pattern. It represents the discrepancies between measured and actual ("true") values of a variable, where the values of the measured variable are equally likely to be higher or lower than the actual values because they are caused by chance fluctuations in measurement. Random-measurement errors are caused by flaws in the measurement process and they tend to cancel each other out and average to zero; however, they increase the variability of measured values. Random-measurement error is also known as "unsystematic" measurement error. Building on our previous example, suppose the researcher realizes her mistake and repeats the ADHD intervention study with two classes that both have recess in the morning, and the measurements are taken immediately following the recess periods. The school counselor is recruited to gather the measurements. Unfortunately, during some of the observation sessions, the school counselor is very alert and attentive, noting each and every disruption among the students. Other times she is distracted by other tasks, sometimes even bringing other work into the observation room with her. On those days, only a portion of the class disruptions are noted on the observation sheets. This creates random-measurement error, because there is no definite pattern of error, but depends solely on the circumstances of the observer. Clearly, either type of measurement error can lead to invalid conclusions regarding the relationships between variables in the study.

Sampling bias. Sampling bias is the intended or unintended systematic distortion of a sample. Sampling bias results in a lack of representativeness of the sample to the population from which it is drawn. The most effective way to avoid sampling bias is to use a random-sampling technique to select your research sample. **Random assignment** is also used in experimental research designs to reduce sampling bias. Random assignment means that the elements in our sample are randomly assigned to the different groups within the research study, for example, the experimental and the control groups. As mentioned previously, random assignment is used in order to strengthen a study's internal validity. Finally, the size of our sample affects the amount of sampling bias present in a research study. The larger the sample size, the more closely it will represent the population on all known variables, as well as variables that have not even been identified that may affect the outcome.

Sampling error. Large sample sizes, random sampling, and random assignment are used to reduce the effects of sampling bias as much as possible. But, regardless of the research design and the researcher's efforts to select a completely unbiased and representative sample, there will still be some sampling error. Suppose we have a population of 100 students and we randomly select 30 of the students. The chance that the mean age of the sample will be exactly equal to the mean age of the population is very small. If we draw a sample of 60 students, we might expect the mean age of the sample to be closer to the population mean, but it will probably still not be exactly the same. Sampling error is the tendency for any sample to differ from the population from which it was drawn. It represents the discrepancy between values in a population and values in a sample (i.e., discrepancy between parameters and statistics). The smaller the sample, the larger the expected sampling error. Therefore, any time a relationship is found between two variables within a sample, there is always a chance that this relationship is just a fluke and does not exist in the population. Alternately, if a relationship is not found between variables in a sample, the relationship may in fact exist in the population.

There are two basic methods for determining if a relationship found in a research sample is likely to be the result of sampling error. These are replication and conducting inferential statistical tests. **Replication** is the process of repeating a research study two or more times using the same population but different samples to see if the same conclusions are found. Clearly, if the same relationship is found using three different random samples, the chance that the relationship can be explained by sampling error is much less than if we used only one sample.

Inferential statistical methods allow us to estimate the probability that an observed relationship between two variables within our sample is actually caused by sampling error rather than an actual relationship. If we conclude that the relationship between variables in our sample is a real one and our sample is sufficiently representative of the population of interest, then we can make the inference that the same relationship exists within the population. Inferential statistical techniques, therefore, provide a basis for generalizing our findings to the larger population. The basic logic behind inferential statistics is the focus of the remainder of this chapter.

SAMPLING DISTRIBUTION OF MEANS

The next step in understanding the underlying logic of inferential statistics is to understand the **sampling distribution of means**. Suppose we select a random sample of 30 social workers and calculate their mean income to be $45,000 a year. Now suppose we take four more random samples of 30 social workers each and calculate the mean incomes for each group. Would you expect the mean incomes for these groups to be the same as the first group? Probably not. Suppose the mean incomes for these groups are $38,000, $36,500, $49,000, and $50,500. If we continue to select random groups of 30 social workers each, we will get a variety of mean incomes for these groups. If we look at the distribution of these sample means, you may have guessed that they will begin to follow a normal curve. If we draw a large enough number of samples, a theoretically infinite number of samples, the distribution of these sample means will follow a normal curve and the mean of the sample means will be a close estimate of the true population mean.

Another useful fact is that regardless of whether or not the population is normally distributed and whether or not our sample observations are normally distributed, the means of the samples will approximate a normal distribution. The larger the number of samples, the more closely the distribution of the sample means will approximate a normal distribution.

In addition to having a mean that closely approximates the population mean, the sampling distribution of means also has a standard deviation. The standard deviation of the sample means is smaller than the standard deviation of the population. In other words, the means of our samples will not differ from each other as much as the actual raw values in our population. The standard deviation of the sample means is referred to as the **standard error of the mean** to distinguish it from the sample and population standard deviations.

The standard error of the mean is basically related to two factors: the size of our sample and the standard deviation of our population. Unfortunately, we generally do not know the population standard deviation. Luckily, if we have a sample size of at least 30 elements, the standard deviation of the sample is a good estimate of the population standard deviation. Therefore, we can calculate the standard error of the mean with only the sample size and sample standard deviation. Surprisingly, the size of the sample is much more significant than the proportion of the population covered by the sample. A sample size of 2,000 provides an accurate estimation of a population whether the population contains 4,000 elements or 4,000,000 elements.

The standard error of the mean will be smaller than the sample standard deviation. How much smaller depends on the size of our sample. The formula for calculating the standard error of the mean (SE) looks like this:

$$\mathrm{SE} = \frac{\textit{Sample standard deviation}}{\sqrt{\textit{Sample size}}} = \frac{sd}{\sqrt{n}}$$

Confidence Interval

When using inferential statistics, we do not actually create a distribution of sample means. The sampling distribution of means is more a theoretical concept than an actual practice. We base our inferences on only one sample with one sample mean and one sample standard deviation. The mean of our sample is viewed as one value in the normal distribution of sample means. We can therefore use the information we have already discovered about normal distributions, mainly the information displayed earlier in Figure 3.6, to infer certain facts about our sample and the population. For example, because we know that our sampling distribution of means is normal, we can assume that our sample mean will fall within one SE on either side of the true population mean 68.26% of the time.

Likewise, we know that our sample mean will fall within three SEs of the true population mean 99.74% of the time.

Looking at it another way, we also find that the true population mean will fall within the range of plus or minus one SE from our sample mean 68.26% of the time. For example, suppose we have test scores for 30 clients, with a sample mean of 60 and a sample standard deviation of 10. The SE would be calculated as follows:

$$SE = \frac{sd}{\sqrt{n}} = \frac{10}{\sqrt{30}} = 10 / 5.48 = 1.82$$

There is a 68.26% probability that the true population mean is within plus or minus one standard error from the sample mean. Therefore, there is a 68.26% chance that the true population mean is between 58.18 (60 – 1.82) and 61.82 (60 + 1.82). This range is called the 68% confidence interval. A **confidence interval** (CI) is a range of values around a given statistic (e.g., the mean) that are believed to contain with a certain probability (e.g., 95%) the true value of that statistic (i.e., the population parameter) (Field, 2009, p. 783). You can also calculate the 95% confidence interval by adding and subtracting two standard errors from the sample mean: 56.36 (60 – 3.64) and 63.64 (60 + 3.64). A 95% confidence interval indicates that there is a 95% probability that any given 95% confidence interval contains that population parameter. Finally the 99.74% confidence interval would be from 54.54 (60 – 5.46) and 65.46 (60 + 5.46), indicating that we can be 99.74% confident that the true population mean falls within this interval. Note that this formula to calculate confidence intervals applies only to normal-shaped distributions (Field, 2009).

Confidence intervals are set *a priori*, usually at either 95% or 99%. They can be displayed in an error bar graph, which shows the range of values (i.e., interval) around the statistic of interest (e.g., the sample mean). The error bar resembles the letter "I," with a point value of a sample statistic plotted somewhere along the "I". On a graph, the error bar starts at the lower CI value, and ends at the upper value. Thus, we can actually see how precise the sample statistic is in terms of reflecting the population parameter. The longer the error bar, the less precise our sample statistic is. Field (2009) provides instructions on how to plot confidence intervals on an error bar graph using SPSS.

HYPOTHESIS TESTING

In the preceding section, we looked at how a theoretical sampling distribution can help us estimate the true population mean. Similar logic can help us determine the probability that an apparent relationship between two variables may be explained by sampling error. This actually involves looking at the values of two variables and determining the probability that these two sets of values came from the same population. For example, suppose a school social worker wanted to determine if there was a difference in absenteeism between male and female high school students. Inferential statistics could compare the absentee rates of the male students and female students to determine the chances that the males and females were drawn from the same population. If there is a high probability that the absentee rates for male and female students are drawn from the same population, then the school social worker would conclude that there is not a significant difference between male and female absenteeism. On the other hand, if there is a high probability that the two sets of rates could not be drawn from the same population, then the school social worker could conclude that there is a significant difference between male and female absenteeism. This process forms the basis for null hypothesis significance testing, and of inferential statistics.

Comparing Two Samples

Previously we looked at how the theoretical sampling distribution of means can be used to estimate the true population mean. Another distribution is used to estimate whether two samples have been drawn from the same or different populations. Not surprisingly, this distribution is the **sampling distribution of the differences between means**. To create this theoretical distribution, let us continue with the previous example. We would draw multiple samples of, say, 30 male high school students and 30 female high school students. For each pair of samples, we calculate the mean absentee rate for the males and the mean absentee rate for the females. Even if there is no difference between the male and female absentee rates, we would not expect the means of these groups to be exactly the same in every case. In some cases, the difference would be positive (male sample mean is larger than female sample mean) and in some cases, the difference would be negative (male sample mean is smaller than female sample mean). Suppose we draw an infinite number of pairs of samples (theoretically of course) and calculate the difference between the means. We could then create a distribution from these differences.

Like the sampling distribution of means, the sampling distribution of the differences between the means would also approximate a normal distribution. If the two samples are from the same population, the mean of the distribution would be 0. In other words, if the samples are from the same population, then the differences between the sample means would center at 0, with progressively fewer of the sample pairs having large positive differences (male sample mean is much higher than the female sample mean) or large negative differences (male sample mean is much lower than the female sample mean). While the mean of this distribution is 0, the variability or standard deviation of this sampling distribution would depend on the standard deviation of the population. While the standard deviation of the sampling distribution of the sample means is called the standard error of the mean, the standard deviation of the sampling distribution of the differences between the means is called the **standard error of the differences between means** (SE-Diff). Because the standard deviation of the population is unknown, we again must estimate the SE-Diff. This is done by combining the standard error of the means (SE) for both of our samples (males and females). The formula for calculating the SE-Diff is as follows:

$$\text{SE-Diff} = \sqrt{(SE_1)^2 + (SE_2)^2}$$

Let us continue our example of female and male absentee rates. Suppose the average rate for the males is 9.5% with a standard deviation of 2.5%, and the average rate for the females is 5% with a standard deviation of 1.5%. We calculate the SEs for both samples as follows:

$$SE_{males} = \frac{sd}{\sqrt{n}} = \frac{2.5}{\sqrt{30}} = \frac{2.5}{5.48} = 0.46$$

$$SE_{females} = \frac{sd}{\sqrt{n}} = \frac{1.5}{\sqrt{30}} = \frac{1.5}{5.48} = 0.27$$

Using these values, we can now calculate the SE-Diff:

$$SE\text{-}Diff = \sqrt{(SE_{males})^2 + (SE_{females})^2} = \sqrt{.46^2 + .27^2}$$
$$= \sqrt{.212 + .073} = \sqrt{.285} = .53$$

The null hypothesis for this example would read:

There is no difference between male and female absentee rates.

A two-tailed research hypothesis would read:

There is a difference between male and female absentee rates.

A one-tailed research hypothesis could be stated in two different directions depending on the school social worker's hypothesis:

The male absentee rate is higher than the female absentee rate.

or:

The female absentee rate is higher than the male absentee rate.

If the null hypothesis is true (the two samples are not significantly different), then the sampling distribution of the differences between the means of pairs of samples would be approximately normal, have a mean of 0, and a standard deviation (or SE-Diff) of about 0.53. Therefore, if there is no difference between these two samples (males and females), the sample means would still differ by up to 0.53 in either direction 68.26% of the time (plus or minus one standard deviation from the mean). The sample means would differ by 1.06 (plus or minus two standard deviations from the mean) 95.44% of the time, and by 1.59 (plus or minus three standard deviations from the mean) 99.74% of the time. In other words, if our two samples do not differ, the two sample means would differ by more than 1.59 only 0.26% of the time.

Rejection Level

The researcher, or school social worker in this example, would decide how much difference between samples would be enough difference to be considered significant before the data are collected. This is referred to as the **rejection level** or significance level, denoted by α (Greek alpha). Suppose the social worker chose a .05 α or rejection level. The rejection level indicates the probability of finding a relationship between variables in the sample as strong as the one observed when, in fact, there is not a real relationship between the variables in the population. In other words, the rejection level is the chance we are willing to accept of rejecting the null hypothesis, when, in fact, the null hypothesis is true. If we set our rejection level at .05, this means the null hypothesis will be rejected if there is less than a 5% chance that the relationship between the two variables can be explained by sampling error.

In our example, the difference between the two sample means is 4.5% (9.5% − 5%). Because the SE-Diff of our sampling distribution is .53, this means that the difference between the means of our two samples is approximately 8 ½ standard deviations from the mean (4.5/.53 = 8.5). Looking back at Figure 3.6, we can see that a difference this large between means of two samples drawn from the same population could only be explained by sampling error a negligible amount of the time. Because the 5% rejection region for a one-tailed hypothesis starts at 1.645 standard deviations (see Figure 3.9), this finding is well within the rejection region. Therefore, if the rejection level is set at .05 or 5%, the social worker would reject the null hypothesis and conclude, based on this information, that there is a relationship between gender and absenteeism in the high school.

One-tailed versus two-tailed hypotheses. We discussed earlier the difference between a one-tailed or directional hypothesis and a two-tailed or nondirectional hypothesis. How does this affect the school social worker's decision as to whether gender and absentee rates differ? Let us

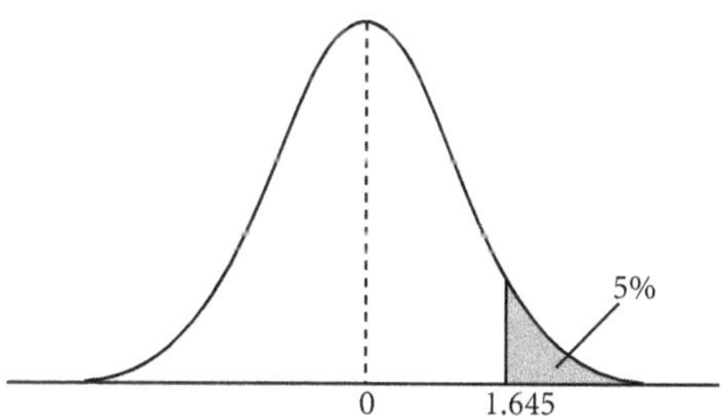

Figure 3.9 Example of a One-Tailed Rejection Region.

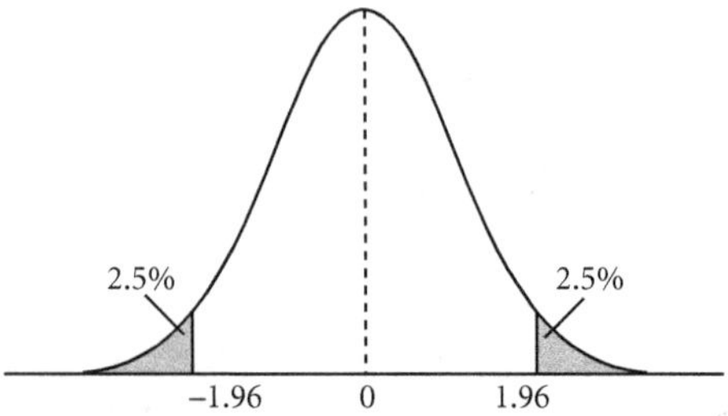

Figure 3.10 Example of a Two-Tailed Rejection Region.

look at this example using two different hypotheses. First, suppose the school social worker stated her research hypothesis that:

There is a difference between male and female absentee rates.

This is a two-tailed or nondirectional hypothesis. Looking at Figure 3.10, you will see that there is a rejection region that lies at both the extreme right and extreme left ends of the graph. Because the rejection region has been set at 5%, this percentage must be split between the right end and the left end of the distribution. There is a 2.5% rejection region represented at the far left end of the graph, and a 2.5% rejection region at the far right of the graph. The rejection region on the right side starts at 1.96 standard deviations and the rejection region on the left side starts at −1.96 standard deviations. In our example, the male sample mean was significantly larger than the female sample mean so our difference lay at the far right end of the graph in the rejection region. If the female sample mean had been significantly larger than the male sample mean, the difference would have fallen in the rejection region at the far left end of the graph, and the null hypothesis would have still been rejected.

Significance Testing

The statistical tests described throughout the rest of this book identify the probability (p value) that a relationship found between two variables may have been produced by sampling error. The ***p* value** indicates the probability of finding a relationship between variables in the sample as strong as the one observed when, in fact, there is not a real relationship between the variables in the population. The p value ranges from 0.00

to 1.00. A *p* value of 0.00 indicates that, assuming the null hypothesis is true, there would be no chance that the relationship observed could be explained by sampling error. A *p* value of 1.00 indicates that the relationship could definitely be explained by sampling error. You would likely never see a *p* value at either extreme, but theoretically it is possible. The closer the *p* value is to 1.00, the more likely sampling error can explain the relationship. Alternately, the closer the *p* value is to 0.00, the less likely the relationship can be explained by sampling error, and the more likely there is a true relationship between the variables.

If the *p* value is *small enough*, the researcher will reject the null hypothesis that there is no relationship between variables and support the research hypothesis that there is a relationship between the variables. The researcher must decide what is small enough. Many researchers agree that if we are 95% sure that the relationship is not caused by sampling error, it is reasonable to conclude that a relationship between the variables probably exists. Remember, we can never completely rule out the chance that sampling error has caused the relationship. A *p* value less than or equal to 0.05 indicates that we are at least 95% certain the relationship is a real one. This is referred to as a .05 rejection level. When researchers want to be more confident that sampling error cannot explain the relationship, they may select a rejection level of .01. They can then be 99% certain that the relationship actually exists in the population.

In some software packages, such as SPSS, the calculated *p* value corresponds only to a nondirectional or two-tailed hypothesis. If the research hypothesis is directional or one-tailed, first the researcher must check to see if the difference is in the direction that he or she predicted. If it is, then the *p* value given is simply divided by two to indicate the chance that the relationship can be explained by sampling error. If the *p* value is less than the rejection level selected, the researcher rejects the null hypothesis and concludes that the relationship predicted in the research hypothesis is statistically significant.

Cummings (2012) provides a caution with regard to interpreting the *p* value in null hypothesis significance testing. He suggests that the correct interpretation of the *p* value is "the probability of getting our observed result, or a more extreme result, *if the null hypothesis were true*" (p. 27). Note that *p* is interpreted based on the assumption that the null hypothesis is true (i.e., there is no difference or relationship between the variables). Note also that *p* does *not* provide "the probability that the null is true" (p. 27). According to Cummings, another misuse of p is to interpret its value on what he refers to as a "sliding scale of degrees of statistical significance" (p. 29), such that the smaller the p value the stronger the relationship or effect. The p value simply tells us whether there is a relationship between two variables; it *does not* indicate the strength of the relationship. The strength of a relationship between two variables is provided by an effect size measure, which is defined later in this chapter and then described in subsequent chapters as it relates to interpreting the findings of the different statistical tests.

Type I and Type II Errors

If the p value is within the rejection level (i.e., $p \leq .05$), does that mean there is definitely a relationship between the variables? No, it does not. It means that we can be at least 95% confident that there is a relationship. We may be incorrect. For example, if the p value is .03, there is still a 3% chance that we will reject the null hypothesis and conclude that the relationship between the variables is a real one when in fact there is not a real relationship between the variables. This is called a **Type I error**. When the p value is greater than our selected rejection level (i.e., $p > .05$), we fail to reject the null hypothesis and conclude that there is not a real relationship between the variables. However, if the p value is .10, then there is a 10% chance that we will fail to reject the null hypothesis and conclude that there is not a real relationship between the variables when in fact there is. This is called a **Type II error**. These errors are not only caused by sampling error. They can be caused by measurement bias, design flaws, or using the wrong statistical test, among others things.

The researcher does have some control over the possibility of committing these errors. As we noted earlier, the researcher selects the rejection level. When we set the rejection region at .05, it means that we are willing to tolerate making a Type I error up to 5% of the time. We can reduce our rejection level to .01 to reduce our acceptance of making a Type I error. However, when we do this, we increase the chance that a Type II error will be committed. It is up to the researcher to decide whether committing a Type I error (concluding that there is a relationship when there is not) or a Type II error (concluding that there is not a relationship when there is) would be more detrimental to the research or dangerous to the

Table 3.1 Hypothesis Decision Table

	Reject Null Hypothesis	*Do not Reject Null Hypothesis*
Relationship between variables is a real one	Conclusion is correct	Type II error
Relationship between variables is not a real one	Type I error	Conclusion is correct

participants. Based on this, he or she can identify the rejection level to be used. See Table 3.1 for a summary of Type I and Type II errors.

One way we can simultaneously reduce our chances of committing a Type I or Type II error is to increase our sample size. The larger our sample, the closer our sample will represent the larger population. This means there will be less sampling error. With less sampling error, we increase the probability that a relationship found in our sample can allow us to accurately predict that the same relationship will be found in the population.

Statistical Power Analysis and Effect Size

More emphasis has been placed in the literature recently on avoiding Type II errors. Researchers have realized that missing relationships between variables that may in fact be significant ones may cause us to reject helpful treatment strategies or ignore variables that might help us predict client problems. The **power** of a statistical test refers to the chance that the test will avoid making a Type II error. Statistical power is the probability that the researcher will reject the null hypothesis when, in fact, it is false. In other words, the researcher will conclude, based on the *p* value, that there is a relationship between variables when there actually is a relationship. Power can be increased by increasing the rejection level, but this will also increase the chance of making a Type I error. Increasing the sample size is generally considered the easiest way to boost the statistical power of a test. A power analysis should be conducted during the design stage of a research study to determine the minimum sample size needed to keep the probability of committing a Type II error to an acceptable level

One condition that relates to the power of the statistical analysis of a sample of data is the actual strength of the relationship that exists in the population. Of course, this is not something the researcher has any control over. The strength of this relationship, however, can be estimated.

This estimation is called the **effect size**. While the p value tells us whether the IV has an effect on the DV, the effect-size (ES) estimates the amount of influence the IV has on the DV. It is an indicator of the strength (i.e., magnitude) and (typically) direction of relationship between the variables. It is clearly informative, beyond information provided by the p value. Thus, there has been increasing interest in and emphasis on effect-size measures. In fact, the APA Manual (2010) recommends reporting effect sizes in inferential statistical testing. It is conventional to report the confidence interval (CI) usually at either 95% or 99%, along with an effect-size measure. Make sure to distinguish between the true ES (i.e., the unknown population parameter) and the observed ES from the sample. To learn more about effect sizes, see Cumming (2012) or Grissom and Kim (2012).

The following is the formula to estimate the effect size in a research study that compares outcomes of an experimental group to a control group to evaluate the effectiveness of an intervention:

$$\textit{Effect size} = \frac{\text{mean for the experimental group} - \text{mean for the control group}}{\text{standard deviation for the control group}}$$

For example, suppose we had 40 clients randomly assigned to an experimental group that received behavioral treatment plus medication to treat obsessive–compulsive disorder (OCD). Another group of 40 clients served as the control group, receiving only medication to treat their disorder. The mean posttest score on the OCD symptom checklist (higher scores represent more severe OCD symptoms) for the experimental group was 40, and the mean score for the control group was 50, with a standard deviation of 5. Suppose an inferential statistical test was performed and it was found that the difference between these mean scores was statistically significant using a .05 rejection level (i.e., $p < .05$).

This tells us that the difference between the groups is large enough that it is unlikely to be the cause of sampling error. However, it does not tell us how much effect the treatment had on the OCD symptoms. To calculate the effect size for this example, the following formula is used:

$$ES = \frac{\text{mean}_{\text{exp}} - \text{mean}_{\text{control}}}{\text{sd}_{\text{control}}} = \frac{40-50}{5} = \frac{-10}{5} = -2$$

Therefore, the mean for the experimental group was two standard deviations below the mean for the control group. Looking back at Figure 3.6, remember a score that is two standard deviations below the mean is lower than 97.72% (50 + 47.72) of the values in the distribution. Therefore, we can conclude that the mean score of the people in the experimental group was lower than 97.72% of the scores of the people in the control group. Another strength of effect size is that, much like the similar *z* score, effect sizes from different studies can be calculated and compared even when different outcome measures and research designs are used.

Statistically Significant versus Meaningful Relationships

When the researcher finds a statistically significant relationship between two variables, it means that it is unlikely that the relationship could have been caused by sampling error. In other words, the relationship is likely a real one. However, this does not necessarily mean it is a meaningful or even an interesting relationship. For example, suppose 800 high school seniors were randomly assigned to complete an SAT preparation workshop before taking the SAT test. Another 600 students were randomly assigned to simply prepare on their own using an SAT preparation manual. After completing their preparation and taking the SAT, the mean SAT score for the first group was 1240 and the mean score for the second group was 1220. Based on the large sample size, this 20-point difference may be statistically significant. But if the workshop costs the school system $200 per student and the preparation manual costs $20, would a difference of 20 points be considered a meaningful enough difference to warrant the financial expenditure? Maybe not. Therefore, the researcher must also consider the practical significance of the findings.

In this chapter, we have presented a foundation for understanding how and why inferential statistics can be used to determine if there is a relationship between the variables of interest in a research study. In the following chapters, we will present the inferential statistical methods most commonly used in social work research.

4

Bivariate Statistical Methods

INTRODUCTION

In the previous chapter, we looked at the concept of statistical inference and how it is used to test research hypotheses. The research hypothesis is a predictive statement regarding the relationship between two or more variables. In this chapter, we will review bivariate statistical methods used to determine the probability that a relationship found between two variables is based on sampling error. There are many bivariate methods, more than we can cover in this book. We will focus on the methods most commonly used in social work research, including the chi-square test, one-sample *t*-test, dependent-samples *t*-test, independent-samples *t*-test, and linear correlation. It is important to understand how researchers select the appropriate statistical test to use in a given research study. Weinbach and Grinnell (2010) point out the five considerations that most directly influence this choice: 1) the sampling method used; 2) the distribution of the variables within the population; 3) the level of measurement of the independent and dependent variables; 4) the amount of statistical power that is desirable; and 5) the robustness of the tests being considered (p. 126).

PARAMETRIC AND NONPARAMETRIC TESTS

There are two basic types of statistical tests: **parametric** and **nonparametric**. The key difference between parametric and nonparametric tests is that, for the most part, nonparametric tests do not require assumptions about the population (i.e., parametric assumptions). Parametric tests have greater power (e.g., the ability to detect small but nonrandom differences) than their nonparametric alternatives. In other words, they are more likely to detect a relationship between two variables when there is, in fact, a relationship. In order to use a parametric test, four assumptions must be met:

- At least one variable (usually the DV) is measured at the interval or ratio level.
- The shape of the distribution of the DV should approach the shape of a normal curve (i.e., unimodal, not skewed, and not kurtotic).

The basis of hypothesis testing is that the data are normally distributed. Thus, nonnormal distributions affect our ability to test hypotheses accurately. Generally nonnormal distributions occur when there are extreme observations (i.e., outliers or influential cases) in the data set. These extreme observations can lead to inaccurate results of our analyses. Violations of the normality assumption can be detected graphically by creating a histogram, frequency distribution, or boxplot of the observations, and then examining its symmetry. The companion website provides step-by-step instructions for assessing violations of the normality assumption with a histogram. Skewness and kurtosis statistics also provide information about the normality of a distribution. In a normal distribution, these values are 0. Thus, the further away they are from 0, the more likely the distribution deviates from normality.

- If two samples drawn from two different populations are being compared, the variances of these different populations should be at least approximately equal.

This assumption is referred to as **homogeneity of variance** or **homoscedasticity.** A violation of this assumption (i.e., unequal variances) is referred to as **heteroscedasticity**. When this assumption is violated in multiple regression analysis, the estimated regression coefficients are not necessarily affected. However, violations may result in inflated standard errors, and thus an increased likelihood of making a Type II error (failing to reject a false null hypothesis and discarding an IV that is actually significant).

The Levene test for homogeneity of variance can be used to evaluate whether the homogeneity of variance assumption has been violated. If the test statistic is nonsignificant, then we retain the null hypothesis that the difference between the variance equals 0, and the assumption has been met. Note that the Levene test is affected by sample size such that, in large samples, it can be significant even when variances are not different. In this case, we recommend calculating the variance ratio. SPSS offers the Hartley's F_{MAX} procedure.

What can be done to correct for violations of this assumption? Minor violations have a negligible impact on the results. To address serious violations, however, one can consider variance-stabilizing transformations, such as taking the log or square root of Y (Lomax, 1992). Another option is to use a different method of estimation that does not require homogeneity of variance, such as the weighted least-square method.

- The observations are independent of one another.

This assumption means that the scores or observations of the DV should not be correlated with one another. As an example, studies that investigate educational questions among children across different classrooms may be at risk of violating this assumption because children in one classroom are likely to have similar experiences with one another (and thus have scores that are correlated with one another), relative to children in other classrooms. This assumption is also more likely to be violated in longitudinal panel studies, when multiple observations are made on the same individuals over time. Violations of this assumption require the use of different methods of analysis (e.g., hierarchical linear modeling). Note that this assumption does not apply to statistical tests (e.g., dependent-samples *t*-test, repeated measures ANOVA) associated with research designs (e.g., experimental designs, panel studies) that are based on data collected at more than one time point from the same sample.

Some parametric tests have additional assumptions, and we may be able to use some parametric tests even when one or more of these assumptions are not met. However, before using any parametric test, it is a good practice to ensure that the data being examined meet the specific assumptions of that test.

If the assumptions for parametric tests are not satisfactorily met, then nonparametric tests may be used. Nonparametric tests have fewer assumptions regarding data characteristics. For example, they can be used with variables measured at the nominal or ordinal level, and the dependent variable does not have to be normally distributed. There are also nonparametric tests that can be used in studies with very small sample sizes.

Selection of the statistical test or tests to use requires that the researcher consider the purpose of the research study, as well as the characteristics of the data being analyzed. Factors to be considered include sampling methods used, sample size, the level of measurement of each variable, distribution of the variables, and possible confounding variables. You must also consider the importance of avoiding Type I and Type II errors for your specific research question. All of these issues help determine the most appropriate statistical test to use for a given research study.

GRAPHICAL REPRESENTATION

Cross-Tabulation Tables

The most common method for graphically displaying the relationship between two nominal- or ordinal-level variables is the **cross-tabulation table.** In a cross-tabulation table, the categories of one variable are displayed as rows in the table, and the categories of the second variable are displayed as columns. If one of the variables is considered the IV and the other is the DV, then generally the categories of the DV are displayed in the columns. However, many researchers present categories of the DV in the rows and categories of the IV in the columns, so it is important to double check this. Both frequencies and percentages are often included, as well as row totals and column totals.

Look at the cross-tabulation table shown in Table 4.1. What types of questions can we answer about the relationship between the two variables *college class* and *GPA?* For instance, one might wonder whether

Table 4.1 College Class by GPA

	GPA				
College Class	*<1.0*	*1.0–1.9*	*2.0–2.9*	*3.0–4.0*	*Total*
Freshman	30 (11%)	75 (30%)	105 (42%)	42 (17%)	252
Sophomore	16 (7%)	67 (29%)	120 (51%)	31 (13%)	234
Junior	8 (4%)	55 (25%)	98 (45%)	57 (26%)	218
Senior	6 (3%)	42 (20%)	88 (42%)	73 (35%)	209
Total	60 (7%)	239 (26%)	411 (45%)	203 (22%)	913

GPAs increase or decrease as students move to the upper level classes. There is a lot of information provided in a cross-tabulation table. Let's examine the rows and columns. What do the row totals (the last column on the right) and the column totals (the last row of figures) tell you? The percentages in this table are called row percentages. In other words, 11% of all freshmen (the top row of figures) have a GPA under 1.0. If you add all of the percentages across the top row (excluding the cell showing the total number of freshmen), they will add up to 100%. You can also use column percentages if they more closely meet your needs. Sometimes, you will find both row and column percentages in a cross-tabulation table. In these cases, there will be two percentages displayed in each cell of the table, one indicating the percentage of the row total and the other indicating the percentage of the column total.

We focus on bivariate statistical methods in this chapter, but we want to mention that cross-tabulation tables can also be used to display the relationship between more than two variables. Suppose we were interested in the pass/fail rates of our MSW students on a licensure test. We are interested in comparing our full-time students versus our part-time evening students, as well as those that entered as "advanced standing" versus those who did not. Look at Table 4.2 for an example of how a cross-tabulation table would display these data. Note in this table, we have chosen to display percentages by column rather than by row, because we are interested in the percentage of each group of students who passed or failed the test. By looking at these data, would you think that these four groups of students differed significantly on pass/fail rates? Probably not. Later in this chapter we will look at specific bivariate statistical methods that are used to determine whether differences such as these are statistically significant.

Table 4.2 Pass/Fail rates for Part-time/Full-time and Advanced Standing versus Regular Students

	Advanced Standing		*Regular Students*		
	Full-time	*Part-time*	*Full-time*	*Part-time*	*Total*
Pass	118 (93%)	121 (92%)	78 (93%)	88 (94%)	405 (93%)
Fail	9 (7%)	11 (8%)	6 (7%)	6 (6%)	32 (7%)
Total	127	132	84	94	437

Scattergram

The **scattergram** is another method for displaying the relationship between two variables. Unlike the cross-tabulation table, the scattergram displays every value of one variable on the *x*-axis along with every corresponding value of the second variable along the *y*-axis. Once the data are plotted, the pattern of the points can suggest a possible relationship between the two variables. Scattergrams are often used to check whether parametric-test assumptions have been met.

Suppose we are interested in the relationship between depression level and severity of eating disordered behaviors. Generally, you plot the IV (or predictor variable) along the *x*-axis and the DV (or outcome variable) along the *y*-axis. If we are hypothesizing that a client's eating disordered behaviors affect or can predict (independent or predictor variable) their depression level (dependent or outcome variable), then a scattergram showing the relationship between these two variables would probably plot the eating disordered behaviors on the *x*-axis and the depression level on the *y*-axis.

In the scattergram displayed in Figure 4.1, the data indicate the higher the severity of eating disordered behavior, the higher the depression level, suggesting a positive correlation between these two variables. Figure 4.2 indicates the opposite, that the higher the severity of eating disordered behavior, the lower the depression level. This would suggest a negative correlation between the two variables. Finally, Figure 4.3 shows a scattergram that suggests little or no relationship between the two variables. We will look later in the chapter at a statistical test that can be used to determine whether correlations such as the ones shown in these scattergrams are statistically significant.

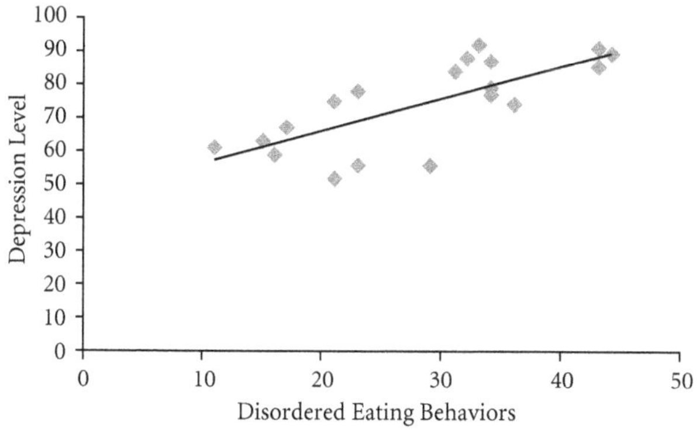

Figure 4.1 Scattergram Suggesting a Positive Correlation.

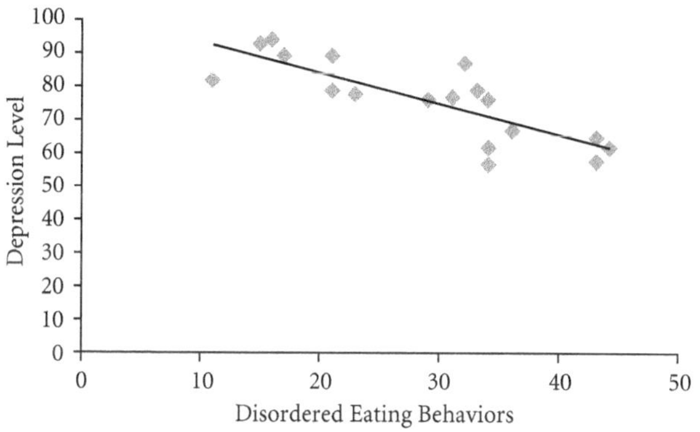

Figure 4.2 Scattergram Suggesting a Negative Correlation.

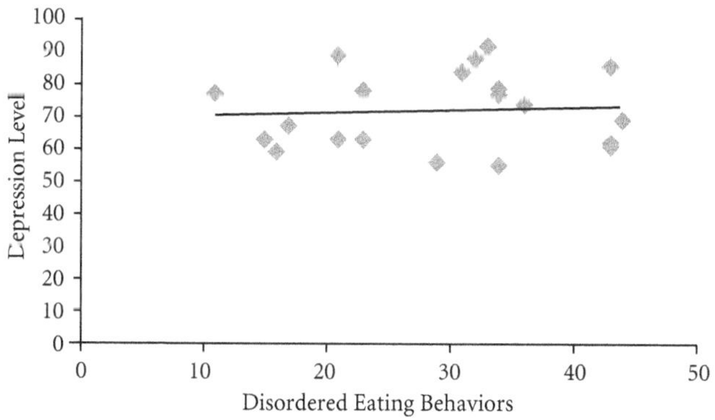

Figure 4.3 Scattergram Suggesting Little or no Correlation.

CHI-SQUARE TEST OF ASSOCIATION

The **chi-square (χ^2) test** measures the association between two variables measured at the nominal or ordinal level. In other words, this test is used with categorical variables only. The chi-square test can be used to test whether there is a relationship between certain nominal level variables and intervention success, and is used extensively in social work research. For example, our IV might be gender (male/female), or group (intervention/control), or type of treatment (cognitive-behavioral/behavioral only/medications), or education level of therapist (bachelors/master's/doctorate), and so forth. The DV, representing treatment success, could be high school graduation (graduated/did not graduate), program completion (completed/did not complete), drug use (used/did not use), etc. The chi-square test tells us if certain value categories of the IV tend to cluster around certain value categories of the DV.

Suppose you want to evaluate a new retention program for high school students who have been deemed at-risk for dropping out of high school before graduation. At-risk students are randomly placed in either the new retention program or they are placed in a control group where they receive the traditional school counseling encouraging them to remain in school. In this example, the IV is the group in which the student is placed (retention program versus control group), and the DV is high school graduation (graduated/did not graduate). The research hypothesis for this study could be stated:

> *The students participating in the new retention program are more likely to graduate high school than those students who receive traditional counseling at the school.*

Our corresponding null hypothesis would read:

> *There is no relationship between participation in the retention program (versus traditional school counseling) and high school graduation.*

The chi-square test would determine if students attending the new retention program were more likely to graduate than the control students. As you will see throughout this chapter and the ones that follow, each of the statistical tests that we describe offer the value of the test statistic, as well as a *p* value. As you will remember from the previous chapter, the *p* value indicates the chance that the relationship between the variables can be explained by sampling error. If we find that 84% of the students in the retention group graduate, while only 64% of the control students graduate, we might conclude that the program was successful. But, depending on our sample size, a difference of 20% might not be statistically significant. In other words, sampling error may be able to account for this difference. For example, say we had 30 students in the retention group and 30 students in the control group, then 25 (85%) from the retention group and 19 (64%) from the control group subsequently graduated. For our purposes, this may seem successful, but statistically the chance of this difference occurring because of sampling error (differences between the two groups that are unaccounted for) may be relatively high, say .20. That is, statistically there is a 20% chance that this difference could be explained by sampling error. Suppose instead there are 300 students in each group, where 252 of 300 students (85%) in the retention group graduated and 192 of 300 students (64%) in the control group graduated. Even though the percentages of graduates in each group remain the same, this difference would almost certainly be statistically significant based on the larger sample sizes in the two groups. In other words, as the sample size increases, the sampling error decreases. Therefore, a smaller difference between the two groups will be statistically significant.

Observed and Expected Frequencies

Several steps are involved in calculating a chi-square statistic. The first step is to create a cross-tabulation table of the data. Table 4.3 shows the frequencies and percentages of the students by group. These are the actual or **observed frequencies**. At this point, you must determine if the frequency in each cell of the table is large enough to meet the requirements of the chi-square test. The general rule of thumb is that

Table 4.3 High School Graduation by Group

Group	*Graduated*	*Did not graduate*	*Total*
Retention Group	48 (84%)	9 (16%)	57
Control Group	32 (64%)	18 (36%)	50
Total	80	27	107

there must be 5 or more in each cell of a chi-square test in which both variables have two categories (2 × 2 table) and 5 or more in 80% of the cells in larger tables. There can be no cells with a frequency of 0. Note we included the row percentages in our cross-tabulation table because we are interested in the percentage of students in the retention group that graduated, rather than the percentage of graduates who were in the retention group. Each time you create a cross-tabulation, you must ask yourself if you are interested in row percentages, column percentages, or both.

The next step is to calculate the **expected frequencies**. These are the frequencies we would expect to see if the null hypothesis was true; in other words, if participating in the retention program made no difference to the graduation rate of the students. An expected frequency is calculated for each cell in the cross-tabulation table by multiplying the row total for the cell by the column total for the cell and then dividing by the grand total or the total sample size (n). The formula looks like this:

$$\text{E(expected)} = \frac{\text{R(ow total)} \times \text{C(olumn total)}}{\text{n(sample size)}}$$

To calculate the expected frequencies for the example shown in Table 4.3, let's start with the cell in the first row and first column. The formula would look like this:

$$\text{E}(1,1) = \frac{\text{Row 1 total} \times \text{Column 1 total}}{\text{Total sample size}}$$

$$= \frac{57 \times 80}{107} = \frac{4{,}560}{107} = 43$$

The calculations for the other three cells would look like this:

$$\text{E}(1,2) = \frac{57 \times 27}{107} = \frac{1{,}539}{107} = 14$$

$$\text{E}(2,1) = \frac{50 \times 80}{107} = \frac{4{,}000}{107} = 37$$

Table 4.4 Observed and Expected Frequencies

	Graduated		*Did not Graduate*		
Group	*Observed*	*Expected*	*Observed*	*Expected*	*Total*
Retention Group	48	43	9	14	57
Control Group	32	37	18	13	50
Total	80		27		107

$$E(2,2) = \frac{50 \times 27}{107} = \frac{1,350}{107} = 13$$

See Table 4.4 for an example cross-tabulation table showing both observed and expected frequencies. With these figures in place, you can now look at the table and compare the observed frequencies with the frequencies that would be expected if graduation outcomes were not associated with participation in the retention program. Note the row totals and the column totals are the same for both the expected and observed frequencies. Looking at the table, do you think the students participating in the retention program were significantly more likely to graduate than those students who did not? The next step is to calculate the chi-square statistic and the associated p value so we can determine if the association between these two variables is statistically significant.

Calculating the Chi-Square Statistic

The chi-square statistic looks at differences between the expected and observed frequencies in each cell. The formula for calculating the chi-square statistic looks like this:

$$X^2 = \sum \frac{(O-E)^2}{E}$$

Here is the calculation for the chi-square value for our previous example:

$$\chi^2 = \frac{(48-43)^2}{43} + \frac{(9-14)^2}{14} + \frac{(32-37)^2}{37} + \frac{(18-13)^2}{13}$$

$$= \frac{25}{43} + \frac{25}{14} + \frac{25}{37} + \frac{25}{13} = .58 + 1.79 + .68 + 1.92 = 4.97$$

You can see from the way this statistic is calculated that the chi-square value is affected by the differences between the expected and observed frequencies. The larger the differences between the expected and observed frequencies, the larger the chi-square value will be. You can also see, however, that the chi-square value is affected by the number of cells or the number of categories in each of the two variables. The more cells in the table, the higher you can expect the chi-square value to be, regardless of whether the difference between the observed and frequencies are large or small. This is true simply because there are more values being added together to determine the chi-square value. Therefore, the chi-square value must be evaluated based on the number of categories in each of the variables.

This is where we will introduce **degrees of freedom**. In most statistical tests, the degrees of freedom must be calculated and taken into consideration when evaluating the test outcome. Degrees of freedom are defined in statistical terms as the number of values involved in the final calculation of a statistic that are free to vary. The degrees of freedom for the chi-square statistic is equal to the number of rows (or number of categories in one variable) minus 1 times the number of columns (or number of categories in the other variable) minus 1. This gives us the number of frequencies we would have to know before we could determine the frequencies for the rest of the cells. For a 2×2 table such as the one in our example, the degrees of freedom would be $(2 - 1) \times (2 - 1) = 1$. In a 3×3 table, the degrees of freedom would be $(3 - 1) \times (3 - 1) = 4$. In other statistical tests that use individual values rather than categorical frequencies, the degrees of freedom are calculated based on sample size (n) rather than the number of categories.

Determining the *p* Value

Once you know the expected and observed frequencies and the degrees of freedom, you can calculate the p value associated with your chi-square value. The p value that is associated with the chi-square test indicates the probability that the differences between the expected and observed frequencies could be the result of sampling error. You calculate the p value by looking at a table of critical values for chi-square (go to the website http://www.statsoft.com/textbook/distribution-tables/for an example chi-square table). Remember that our example is based on a one-tailed research hypothesis (*The students participating in the new retention program are more likely to graduate high school than those students who receive traditional counseling at the school*). Our chi-square value is 4.97 and the degree of freedom is 1. Now we can go to the table to determine the *p* value. Look at the first column on the left and locate the degrees of freedom (*df*). Because it is 1, we need to look at the values on the first line. Continue across the line until you find the value that is larger than our chi-square value (4.97). Our value falls between 3.84 and 5.02. To determine the *p* value, we need to look across the top row of *p* values. Our chi-square value falls between the value corresponding to $p = .025$ and $p = .05$, therefore we know the actual *p* values lies between .025 and .05. The way this is presented is $p < .05$. In other words, you select the larger of the two *p* values and state that the *p* value for your test is less than that *p* value. If the chi-square value had been larger than the last value in the row for $df = 1$ (7.88), we would have presented the *p* value as $p < .005$. On the other hand, if the chi-square value had been smaller than the first value in the row for $df = 1$ (.00004), we would have presented the *p* value as $p > .995$.

Interpreting and Presenting the Results

Suppose in our example that we have chosen the traditional rejection level $\alpha = .05$. Based on our findings, can we reject the null hypothesis? Before we make this decision, we must consider one more thing. The chi-square test determines the probability that the relationship between the two variables could be caused by sampling error. In our example, this probability is less than .05 suggesting that we can indeed reject the null hypothesis. But, the chi-square statistic only looks at the difference between expected and observed. It does not consider the direction of the difference. In other words, because we are testing a one-tailed or directional hypothesis, we must check to make sure the difference is in the desired direction. The chi-square test would show the same relationship whether the students in the retention groups had a higher graduation rate or a lower graduation rate than the students who received the traditional services. Looking at the data, we can see that the students in the retention groups had a higher graduation rate than the control students; therefore, the differences identified by the chi-square statistic are in the desired direction. We can, therefore, safely reject the null hypothesis based on a p value of less than .05. In cases where a two-tailed or nondirectional research hypothesis is being tested, this last step can be omitted because any statistically significant difference found would support our research hypothesis.

When presenting the results from a chi-square statistical analysis, three values must be presented: the chi-square value itself (χ^2), degrees of freedom (df), and the probability that sampling error could have produced the differences between observed and expected frequencies (p). In our example, the results would be presented as follows:

$$\chi^2 = 4.97, df = 1, p < .05$$

Another way these results can be represented is:

$$[\chi^2_{(1)} = 4.97; p < .05]$$

Examples of Chi-Square Test in the Literature

Use of the chi-square test can be found throughout the social service literature. Chapman, Desai, Falzer, and Borum (2006) used chi-square analyses to examine the risk and protective factors of 757 juvenile offenders admitted to youth detention centers. A chi-square test revealed that African American youths were more likely than White youths to be detained for serious offenses that were transferred automatically to adult court. While African American youths were more likely to come from disorganized, highly violent neighborhoods, they were less likely to score high on the drug or alcohol use scales, and less likely to show a high risk of violence. These were partially explained by the higher level of protective factors found among the African American youths, including stronger attachments and bonds and more engagement in socially acceptable pursuits.

The chi-square test is also used to determine if a research sample is representative of the larger population with regard to some characteristic that is measured at the nominal or ordinal level. For example, in this same study, Chapman et al. (2006) noted that the distribution of race in the youth offenders was not representative of race in the population of the state in which the study was conducted, with African American and Hispanic youths overrepresented and White youths underrepresented in the detention center.

In another study (McClatchey, Vonk, & Palardy, 2009), the effectiveness of a camp-based, trauma-focused grief intervention was examined. The sample included 100 children who had lost a parent, and the treatment focused on reducing traumatic grief and posttraumatic stress disorder symptoms among the children. The results from the chi-square test revealed that children who did not attend the camp were much more likely to continue experiencing severe PTSD symptoms than children who did attend the camp. In addition, the chances of experiencing severe grief were much greater for children who did not attend the camp.

Chi-square analysis is also used to compare an experimental group with the control or comparison group to determine whether the groups are equivalent on characteristics that have been deemed important by the researcher. In this study (McClatchey et al., 2009), chi-square tests were used to ascertain that the two groups were equivalent with regard to gender, race, age group, type of loss (mother or father), and type of death (expected or unexpected).

In addition to comparing two groups based on whether the participants completed a treatment (experimental group) or not (comparison group), the chi-square test is also used to compare variables of interest based on the attributes of a dichotomous variable. For example, Myers (2010) looked at drug, alcohol, and tobacco use, and use of violence among adolescents in a rural, southern community, based on race (Caucasian versus African American), gender (male versus female), and age (middle school–aged children versus high school–aged children). She studied more than 2,500 middle and high school youth, and concluded that Caucasian youth were more likely to use tobacco and alcohol, while African American students were more likely to engage in violent activities. Male students were more likely to use tobacco and drugs than female students, and were more likely to engage in violent activities. Finally, older students were more likely to use tobacco, alcohol, and drugs, and to participate in violent behaviors than younger students.

Using SPSS

In this demonstration we explain how to conduct the chi-square test in SPSS, version 19. The Chi-Square test is used to evaluate significant differences between groups, such as differences in diagnosis between males and females. The chi-square test can only be conducted with categorical data, measured at the nominal or ordinal level. In this example, we use two categorical variables from the NELS data set, SEX and DROPOUT_RISK. SEX is an indicator of the sex of the respondent. The value of "1" indicates "male" and the value "2" indicates "female." DROPOUT_RISK indicates the number of risk factors that each respondent has for dropping out of school. The value "0" indicates that the respondent has no risk factors; the value "1" indicates that the respondent has one or more risk factors.

- Open the SPSS data file entitled, BIVARIATE.SAV.
- Click on "Analyze" in the menu bar at the top of the data view screen.
- A drop down menu will appear. Move your cursor to "Descriptive Statistics."
- Another drop down menu will appear to the right of the first menu. Click on "Crosstabs" to open the Crosstabs dialog box.
- Highlight "SEX" in the box on the left side of the Crosstabs dialogue box. Click the arrow to move it to the box entitled, "Row."
- Highlight "DROPOUT_RISK" in the box on the left side of the Crosstabs dialogue box. Click the arrow to move it to the box entitled, "Column."
- Click "Statistics" at the top right of the dialogue box.
- Click "Chi Square" in the upper, left corner of the dialogue box.
- Click "Continue" to close the dialog box.
- Click "OK." SPSS should now calculate the chi square test statistic and produce a cross-tabulation of respondent sex (SEX) by high school dropout risk status (DROPOUT_RISK).
- One difference when using SPSS to conduct the chi-square test rather than using the table of values we looked at earlier is that the p value is presented as an exact value (e.g., $p = .046$ instead of $p < .05$).
- NOTE: A copy of this output is available on the companion website. The file is entitled, "Chi Square test results.spv."

Alternatives to the Chi-square Statistical Test

Technically, the chi-square test can be used for variables measured at the nominal or ordinal levels. However, it is generally agreed that the chi-square test should not be used with ordinal-level data, because the test ignores the ranking of the different categories. Alternative nonparametric tests, such as the Mann–Whitney U, the Median test, and the Kruskal–Wallis test, are alternatives to the chi-square test when considering ordinal-level data (Krysik & Finn, 2010).

Sample size is another consideration when deciding whether to use the chi-square test. In order to use the chi-square test, the sample size must be large enough so that the expected frequency in each cell is at least 5. If your sample size is small, you can combine cells (when appropriate) in order to increase the expected frequencies in each cell. In cases where the frequency for one or more of the cells is less than 5 and the cells cannot be collapsed, another statistical test, the *Fisher exact test*, may be used to determine the relationship between the two variables.

t TESTS

In social work research, the IV is often measured at the nominal level. For example, we may be comparing an experimental group to a control group or a wait-listed comparison group. Or we may want to compare two or more groups based on some demographic characteristic, such as gender, race, or sexual orientation. In cases where the DV is also at the nominal or ordinal level, we would use the chi-square test or one of the alternatives discussed earlier. However, when the DV is measured at the interval or ratio level, which is often the case, we would turn to *t* tests and simple analysis of variance (also referred to as simple ANOVA or one-way ANOVA).

The *t* test is used when the IV is dichotomous (nominal with only two categories) and the DV is measured at the interval or ratio level. When the IV has more than two categories, the simple ANOVA must be used. The *t* test is a parametric statistical test, and therefore should meet the general requirements of parametric procedures. We describe three basic forms of the *t* test in detail: the one-sample *t* test, the dependent-samples *t* test, and the independent-samples *t* test.

One-Sample *t* Test

The one-sample *t* test compares the mean of the DV (interval or ratio level) for the research sample to the known or estimated mean of the population from which the sample was drawn. It is used to determine if the sample is typical of the population with regard to the DV, or if it differs significantly from the population. The known or estimated population mean is called the *test value.* The formula for this test produces a value for *t*, and then the researcher must refer to a table of critical values using the *t* value and the degrees of freedom to find the *p* value. The calculation for degrees of freedom for a one-sample *t* test is the sample size (n) minus 1.

The one-sample *t* test is used for two general purposes. The first is to determine if the sample drawn from a population is representative of the population in relation to some important interval or ratio-level variable. For example, suppose we are conducting a study to determine if gender is related to scores on a standardized instrument measuring self-esteem among students in our high school. We select a sample of 30 students from the high school, and we would like our sample to be representative of the population of all students on the variable *age*, a variable that we suspect might be a confounding variable affecting self-esteem. We would calculate the one-sample *t* test by comparing the mean age of the sample to the mean age of all of the students at the high school. In this case, we would actually hope for a large *p* value. A large *p* value would indicate that the mean age of the sample is not statistically different from the mean age of the population. Therefore, we could continue our study with this sample. A small *p* value (i.e., less than .05) would indicate that the mean age of the sample is statistically different from the mean age of the population, therefore we might draw another sample.

The second purpose of the one-sample *t* test is to determine if a research sample differs from the population on some interval or ratio-level variable to the extent that the difference is not likely caused by sampling error. This same test can be used in situations in which we hope to see a significant difference between our sample and the greater population, such as in cases where our sample has completed some intervention that we hypothesize will significantly change their score on some interval or ratio-level measurement.

For example, suppose you are trying to determine if the students attending your community's magnet program in the arts differ from the larger population on an IQ test. You already know that the average IQ score is 100, so 100 is set as the test value. If you are hypothesizing that these students' IQ scores will be higher than the average population IQ score, your one-tailed research hypothesis would read:

> *The mean IQ score for the students involved in the magnet program is higher than the mean population IQ score of 100.*

The corresponding null hypothesis would read:

> *The mean IQ score for the students involved in the magnet program does not differ from the mean population IQ score of 100.*

Suppose we select 30 students from the magnet program and complete an IQ test with each student. The mean of these IQ scores is 108 and the standard deviation is 6. We now have all the information we need to calculate the value of *t* and determine if there is support for our research hypothesis. The formula for calculating the value of *t* for the one-sample *t* test looks like this:

$$t = \frac{\overline{X} - \mu}{s / \sqrt{n}}$$

where $\overline{X}$ is the mean score for the sample, μ is the mean for the population or test value, *s* is the standard deviation for the sample, and *n* is the sample size. Putting in the values we have:

$$t = \frac{108 - 100}{6/\sqrt{30}} = \frac{8}{6/5.48} = \frac{8}{1.09} = 7.34$$

Therefore, the *t* value is 7.34. Next we must locate the *p* value to determine if the difference is statistically significant or can likely be explained by sampling error.

To find the *p* value, we need to refer to a table showing the *t* distribution (go to the website http://www.statsoft.com/textbook/distribution-tables/ for an example t table). Because the sample size is 30 in this example, the degrees of freedom would be 29 ($n - 1$). Look down the column of degrees of freedom until you reach 29. Then move across the row of figures until you find where 7.34 falls. Our calculated *t* value is larger than the value in the last column, therefore our *p* value is less than .0005.

Because $p < .0005$, there is almost no chance that sampling error can explain the difference between the sample's mean IQ score and the population's mean score. Therefore, we would reject the null hypothesis. The results would be presented as follows:

$$t = 7.34,\ df = 29,\ p < .0005 \text{ or } [t\ (29) = 7.34;\ p < .0005]$$

Examples of one-sample *t* test in the literature. The one-sample *t* test is used in the social science literature to compare data from a research sample to known population data. One study by Kim, Szatmari, Bryson, Streiner, and Wilson (2000) looked at the prevalence of mood and anxiety issues among children with Asperger Syndrome and autism. The researchers used the one-sample *t* test to compare the scores of the children in the sample on several measures with population norms that had already been established for these measures. They concluded that the children with autism and Asperger Syndrome scored significantly higher than the population means on depressive, overanxious, ADHD, and oppositional behaviors. They did not score significantly higher than population means on conduct disorder or separation anxiety.

Kagee (2008) studied 119 patients being treated for diabetes or hypertension to determine if these patients were more likely to experience symptoms of depression and anxiety than the general population. Kagee compared patient scores on measures of anxiety and depression to the cut-point score that had already been established to determine clinical significance, and found that the sample mean did not differ significantly from the cut-point score. The results did, however, suggest that a significant number of patients were experiencing significant psychological difficulties for which they may not have been receiving treatment.

Alternatives to the One-Sample *t* Test. If the DV is measured at the nominal or ordinal level, the chi-square test would be used instead of the one-sample *t* test to accomplish the same thing. The chi-square test would be used, for example, if you wanted to determine if your sample was representative of the population with regard to race, marital status, or some other categorical variable.

Dependent Samples *t* Test

The dependent samples or paired-samples *t* test is used to determine if there is a statistically significant difference between the scores from two samples in which the individual values or cases in the sample are paired with each other for some reason related to the research question being asked. Three examples of when two samples of data may be dependent (i.e., paired) are:

- Individual cases in two different groups are sometimes paired with each other, either naturally (such as twins) or based on some characteristic. Both groups are then measured on some interval or ratio-level variable and the values are compared across each pair in the sample.
- One sample of cases that have been measured on some interval or ratio-level variable two times, such as in a pretest and posttest outcome study.
- In some situations, two different, but related, scores will be compared to each other for one sample of cases (such as comparing reading and writing scores from some standardized test).

In each of these instances, we have a series of paired values to put into the formula. The degrees of freedom for a dependent-samples *t* test is the number of pairs of values minus 1. When one sample is being measured two times, the degrees of freedom are equal to the sample size minus 1. When two groups have been paired up and measured one time, degrees of freedom are equal to the number of pairs of cases minus 1.

Paired Samples Measured One Time. In this situation, two groups or samples have been matched one-to-one based on some characteristic. For example, in twin studies, the twins are paired with one twin being in one group and the second twin being in the second group. In other examples, we may want to match group members on some variable we believe could be a confounding variable to ensure that the influence of the confounding variable is controlled. See Table 4.5 for an example of two samples that have been matched as closely as possible on the variable, *age*, and measured on the DV, *self-esteem level.*

One Sample Measured Two Times. In this situation, one sample is measured two times on an interval or ratio-level DV. Research studies that involve only one sample measured at pretest and posttest would use the dependent-samples *t* test to determine if there was a statistically significant change between the two administrations of the measure. See

Table 4.5 Two Samples Matched on Age

Pair	*Name*	*Age*	*Self Esteem*
1	Joe	22	45
1	Susan	22	32
2	Sandra	32	48
2	Eliot	33	31
3	Sam	28	56
3	William	27	47
4	Rob	36	32
4	Jennifer	36	27

Table 4.6 for an example of one sample measured two times on the DV, *self-esteem level.*

Example. Using the example data in Table 4.6, we will now turn to detailed instructions on how to calculate a t statistic for the dependent-samples t test. Remember that, ideally, this test should be used with a sample size of 30 or more. We are using a smaller sample size simply for convenience in presenting the calculations. The formula for the dependent samples t-test looks like this:

$$t = \frac{\overline{X}_2 - \overline{X}_1}{\sqrt{\dfrac{\Sigma D^2 - \dfrac{(\Sigma D)^2}{n}}{n(n-1)}}}$$

where $\overline{X}_1$ is the mean score for sample one, $\overline{X}_2$ is the mean score for sample two, D refers to the differences between the pairs of scores, and n is the sample size. Suppose the eight people in this example completed a

Table 4.6 Students Scores on Pretest and Posttest

Student	*Name*	*Pre-test*	*Post-test*
1	Joe	27	31
2	Susan	32	38
3	Sandra	38	48
4	Eliot	39	54
5	Sam	22	42
6	William	35	37
7	Rob	31	47
8	Jennifer	37	48

scale measuring self-esteem before and after attending an 8-week cognitive therapy treatment designed to increase self-esteem. You are now trying to determine if the self-esteem level after completing the treatment is significantly higher (improved) than the scores before completing the session. The research hypothesis could read:

The self-esteem levels of the clients after completing the cognitive treatment are higher than the self-esteem levels before treatment.

The null hypothesis would be stated as:

There is no difference between the self-esteem levels before and after completing the cognitive therapy treatment.

To calculate the value of *t*, you must first subtract the pretest score from the posttest score for each case (see Table 4.7 for computation details). The sum of these differences is 84. Then square the differences and calculate the sum of the squared differences. Next calculate the means of the pretest and posttest scores:

$$\overline{X}_1 = \frac{261}{8} = 32.63$$

$$\overline{X}_2 = \frac{345}{8} = 43.13$$

Table 4.7 Computing a *t* Value for One Sample Measured Two Times

Student	*Pre-test Score* *(X_1)*	*Post test Score* *(X_2)*	*Difference* *(D)*	*Difference*^{2} *(D^2)*
1	27	31	4	16
2	32	38	6	36
3	38	48	10	100
4	39	54	15	225
5	22	42	20	400
6	35	37	2	4
7	31	47	16	256
8	37	48	11	121
	$\Sigma X = 261$	$\Sigma Y = 345$	$\Sigma D = 84$	$\Sigma D^2 = 1{,}158$
	$\overline{X}_1 = 261/8 = 32.63$	$\overline{X}_2 = 345/8 = 43.13$		

Following the formula for a dependent-samples *t* test, you have the following calculations:

$$t = \frac{43.13-32.63}{\sqrt{\dfrac{1{,}158-\dfrac{7{,}056}{8}}{8(7)}}} = \frac{10.5}{\sqrt{\dfrac{1{,}158-882}{56}}} = \frac{10.5}{\sqrt{\dfrac{276}{56}}} = \frac{10.5}{\sqrt{4.93}} = \frac{10.5}{2.22} = 4.73$$

The value of *t* is 4.73. Looking at the table of critical values of *t* (go to the website http://www.statsoft.com/textbook/distribution-tables/for an example *t* table), you can locate the *p* value. Degrees of freedom is equal to the number of pairs of values (8) minus 1 ($df = 7$). Look down the column of *df* until you get to 7. Then look across the row of values until you find the largest value that is still smaller than 4.73. In this case, that number is 3.499 (the next number, 5.408, is larger than our *t* value). Looking at the *p* values above these two numbers, we determine that our *p* value falls between .005 and .0005. Selecting the larger of these two *p* values, we conclude that our *p* value is less than.005.

If we set our rejection level at $\alpha < .05$, then we would reject the null hypothesis and conclude that the clients scored statistically significantly higher on the measure of self-esteem after completing the cognitive therapy treatment. These results are presented as follows: $t = 4.73$, $df = 7$, $p < .005$ or $[t\ (7) = 4.73;\ p < .005]$. Note that this simply indicates that the mean pre- and posttest scores are statistically different. This does not indicate the magnitude or the size of the difference. To determine this, a measure of effect size would need to be calculated. See Grissom and Kim (2012) for information on how to calculate an effect-size measure of the difference between two means and accompanying confidence intervals.

Examples of dependent samples *t* test in the literature. The dependent-samples *t* test can be found throughout the social science literature. The most common use is comparing pretest and posttest data. Singh et al. (2007) looked at the effects of mindfulness training for parents of children with autism. Comparing pretest and posttest measures of parental aggressive behaviors, the researchers discovered that the aggressive behaviors were significantly reduced and the children's social skills had increased after mindfulness training was completed. In addition, the parents were more satisfied with their parenting, reported more social interaction with their children, and experienced less parenting stress.

A study by Wang, Blomberg, and Li (2005) provides an example of how the dependent-samples *t* test can be used to test one measure across two samples that have been matched based on some characteristic deemed important by the researcher. In this study, a sample of delinquent youth was paired with nondelinquent youth based on age, race, gender, socioeconomic status, exceptionality status, and school type. The matched pairs were then compared in terms of academic achievement, school attendance, and disciplinary problems. By comparing each delinquent youth to a nondelinquent youth who was equivalent with regard to age, race, and the other control variables, the researchers were able to determine if delinquency and academic deficiencies were related. Wang et al. (2005) used the dependent-samples *t* test to compare 5,187 matched pairs of adolescents, and found that the delinquent youth were more likely to suffer from educational deficiencies than the nondelinquent comparison youth. The delinquent adolescents were more likely to have lower grade point averages and lower attendance rates, had significantly more disciplinary problems, and were more likely to be retained and have to repeat grades in school.

Using SPSS. In this demonstration we explain how to conduct the Paired-Samples *t* Tests in SPSS, version 19. The Paired Samples *t* Test is used to determine if there is a significant difference between the means of two interval or ratio-level variables for a single group, such as a difference between the means on pretest and posttest scores in an experiment. In this example, we use two ratio-level variables from the NELS data set, BY2XRSTD and BY2XMSTD. BY2XRSTD provides a measure of standardized reading scores and BY2XMSTD provides a measure of standardized math scores.

- Open the SPSS data file entitled, BIVARIATE.SAV.
- Click on "Analyze" in the menu bar at the top of the data view screen.
- A drop down menu will appear. Move your cursor to "Compare Means."
- Another drop down menu will appear to the right of the first menu. Click on "Paired Samples T Test" to open the Paired Samples *t* Test dialog box.
- Highlight BY2XRSTD in the box on the left side of the Paired Variables dialogue box. Click the arrow to move it to the Paired Variables box.

- Highlight BY2XMSTD in the box on the left side of the Paired Variables dialogue box. Click the arrow to move it to the Paired Variables box.
- Click "OK." SPSS should now calculate the paired-samples *t*-test statistic.
- NOTE: A copy of this output is available on the companion website. The file is entitled, "Paired-Samples *t*-Test results.spv."

Independent-Samples *t* test

Finally, the independent-samples *t* test compares the mean scores on an interval or ratio-level DV between two independent groups. For example, this test would be used to compare scores between an experimental group and a control group, or between a sample of females and a sample of males. The sample size for the two groups does not have to be equal. The degrees of freedom for the independent-samples *t* test equals the total sample size of the two groups combined minus 2.

Example. Suppose a group of 21 new employees are scheduled to participate in a preparation class before taking an exam to measure their understanding of agency policies. Remember again that we are using a smaller than customary sample size in order to present the example. In an effort to evaluate a new on-line preparation course, the supervisor randomly assigns 10 employees to complete the traditional face-to-face class and 11 employees to complete the new on-line course. The supervisor is confident that the traditional class will be more effective in preparing the employees, because he feels the personal contact and the opportunity to ask questions throughout the presentation are vital to helping the employees understand the policies. He wants data to help him argue that the traditional courses should continue to be taught. He believes that if the outcome of the group completing the on-line course is as good as or better than the group completing the traditional course, the traditional course will be cancelled because the on-line course is cheaper and easier to administer. Therefore, he writes his hypothesis as follows:

The employees who complete the traditional face-to-face preparation course will score higher on the exam than those employees who complete the on-line preparation course.

Table 4.8 Computing a Value for Independent Samples

Exam Scores		*Exam Scores Squared*	
Traditional (X) (N_Y = *10)*	*On-Line (Y)* (N_Y = *11)*	*Traditional* (X^2)	*On-Line* (Y^2)
96	81	9,216	6,561
87	73	7,569	5,329
95	81	9,025	6,561
89	78	7,921	6,084
83	74	6,889	5,476
76	91	5,776	8,281
89	79	7,921	6,241
98	89	9,604	7,921
100	75	10,000	5,625
84	67	7,056	4,489
	71		5,041
$\Sigma X = 897$	$\Sigma Y = 859$	$\Sigma X^2 = 80,977$	$\Sigma Y^2 = 67,609$
$\mu_x = 897/10 = 89.7$	$\mu_y = 859/11 = 78.1$		

The first step is to calculate the mean or average score for both groups (see Table 4.8 for data and calculations). The mean for the traditional course is 89.7 and the mean for the on-line course is 78.1. The independent-samples *t* test can be used to determine if there is a statistically significant difference between these two means, or if sampling error can more plausibly account for this amount of variation. The formula for calculating the value of *t* for two independent samples with different sample sizes and equal variances is:

$$t = \frac{\left|\overline{X}_2 - \overline{X}_1\right|}{\sqrt{\frac{(n_1 - 1)(s_1^2) + (n_2 - 1)(s_2^2)}{n_1 + n_2 - 2}\left(\frac{1}{n_1} + \frac{1}{n_2}\right)}}$$

where $\overline{X}_1$ is the mean of the first sample, $\overline{X}_2$ is the mean of the second sample, n_1 is the sample size for the first sample, n_2 is the sample size for the second sample, s_1 is the standard deviation for the first sample, and s_2 is the standard deviation for the second sample. The next step is to calculate the standard deviations for both groups. The formula for the standard deviation for our samples looks like this:

$$S_1 = \sqrt{\frac{n_1 \Sigma X_1^{\,2} - (\Sigma X_1)^2}{n_1(n_1 - 1)}} \qquad S_2 = \sqrt{\frac{n_2 \Sigma X_2^{\,2} - (\Sigma X_2)^2}{n_2(n_2 - 1)}}$$

Replacing the variables with our values, we have:

$$s_1 = \sqrt{\frac{10(80{,}977) - 897^2}{10(9)}} = \sqrt{\frac{809{,}770 - 804{,}609}{90}} = \sqrt{\frac{5{,}161}{90}} = \sqrt{57.34} = 7.57$$

$$s_2 = \sqrt{\frac{11(67{,}609) - 859^2}{11(10)}} = \sqrt{\frac{743{,}699 - 737{,}881}{110}} = \sqrt{\frac{5{,}818}{110}} = \sqrt{52.89} = 7.27$$

Note that a different form of the standard deviation formula was used in calculating the standard deviations above. The results are identical to the formula used in Chapter 2.

Next, put the means, standard deviations, and samples sizes for both groups into the formula, and our t value can be calculated:

$$t = \frac{|78.1 - 89.7|}{\sqrt{\frac{(10-1)(7.57^2) + (11-1)(7.27^2)}{10+11-2}\left(\frac{1}{10} + \frac{1}{11}\right)}}$$

$$= \frac{|-11.6|}{\sqrt{\frac{(9)(57.3) + (10)(52.9)}{19}(.1 + .09)}} = \frac{11.6}{\sqrt{\frac{515.7 + 529}{19}(.19)}}$$

$$= \frac{11.6}{\sqrt{\frac{1{,}044.7}{19}(.19)}} = \frac{11.6}{\sqrt{54.98(.19)}} = \frac{11.6}{\sqrt{10.45}} = \frac{11.6}{3.23} = 3.59$$

The value of t is 3.59. Looking at the table of critical values of t (go to the website http://www.statsoft.com/textbook/distribution-tables/for an example t table), you can locate the p value. Degrees of freedom are equal to the sum of the two samples sizes (21) minus 2, which equals 19. Look down the column of *df* until you get to 19. Then look across the row of values until you find the largest value that is still smaller than 3.59. In this case, that number is 2.861 (the next number, 3.883, is larger than our *t* value). The *p* value that is directly above 2.861 is .005. Therefore, our *p* value is less than .005.

If we set our rejection level at $\alpha < .05$, then we would reject the null hypothesis and conclude that the employees that completed the in-person class scored statistically significantly higher on the evaluation than the employees that completed the on-line course. Therefore, the supervisor can reject the null hypothesis and argue that the employees learn the policies better when taking the in-person class. This statistic would be presented as follows: $t = 3.59$, $df = 19$, $p < .005$ or [t (19) = 3.59; $p < .005$]. Once again, this indicates that the averages of the two groups are statistically different from one another, but does not indicate the size of the difference. An effect-size measure can be calculated to determine the size or magnitude of the difference.

Examples of independent samples *t* test in the literature. Independent-samples *t* tests are used throughout the research literature to compare variables of interest between experimental and control groups. Lev-Wiesel and Liraz (2007) analyzed the narrative descriptions of negative life events given by 60 children whose fathers were drug abusers to determine if drawing prior to the narratives added richness to the descriptions given. Each child in the experimental group was asked to draw his or her life in the shadow of their father's addiction to drugs. They were then asked to verbally describe their life in the same shadow. The children in the comparison group were asked only to verbally describe their lives. The researchers concluded that the narratives of the children who had first drawn their lives were more detailed and more revealing of emotions than the narratives of those children who had not drawn prior to providing their narratives.

The independent-samples *t* test can also be used to compare a sample across the attributes of a dichotomous variable, such as gender. Rayle et al. (2006) examined the gender differences among American Indian middle school students with regard to frequency of drug offers and difficulty in refusing drugs from various people who offer the drugs. They found that the female students reported more offers of illegal drugs from friends, cousins, and other peers than the male students. The female students also reported more difficulty in responding to and refusing drug offers from all sources.

In addition, like the chi-square test, the independent samples *t* test is also used to determine the equivalence of two groups in a research study. While the chi-square is used to compare groups based on nominal and ordinal-level variables, the *t* test is used to compare groups based on interval and ratio-level variables. For example, before looking at the gender differences, Rayle et al. (2006) first used the independent-samples *t* test to determine that the males were not significantly different from the females with regard to age or grade level. They used the chi-square test to verify equivalency on nominal variables, such as whether or not students lived on a reservation and whether they received free or reduced-priced school lunch.

Using SPSS. In this demonstration we explain how to conduct the Independent Samples *t* Tests in SPSS, version 19. The Independent Samples *t* Test is used to determine whether there is a significant difference between mean scores of two groups, such as a difference between males and females on their mean level of depression. One variable must be dichotomous; the other variable should be measured at the interval or ratio level. In this example, we use SEX for the dichotomous variable and BY2XRSTD, for the ratio level variable. SEX is an indicator of the sex of the respondent. The value of "1" indicates "male" and the value "2" indicates "female." BY2XRSTD provides a measure of standardized reading scores.

- Open the SPSS data file entitled, BIVARIATE.SAV.
- Click on "Analyze" in the menu bar at the top of the data view screen.
- A drop down menu will appear. Move your cursor to "Compare Means."
- Another drop down menu will appear to the right of the first menu. Click on "Independent Samples T Test" to open the Independent Samples T Test dialog box.
- Highlight BY2XRSTD in the box on the left side of the Test Variable(s) dialogue box. Click the top arrow to move it to the Test Variables box.
- Highlight SEX in the box on the left side of the Test Variable(s) dialogue box. Click the bottom arrow to move it to the Grouping Variable box.
- Click on "Define groups."
- Enter "1" (male) in the Group *1* box; Enter "2" (female) in the Group *2* box.

- Click "Continue" to close the dialog box.
- Click "OK." SPSS should now calculate the independent samples t-test statistic.
- NOTE: A copy of this output is available on the companion website. The file is entitled, "Independent-Samples T-Test results. spv."

Bridge to Advanced Statistical Tests

There is a family of statistical procedures that is related to the *t* test. The simple or one-way analysis of variance (simple ANOVA) is used much like the independent-samples *t* test to compare one IV measured at the nominal or ordinal level and one DV measured at the interval or ratio level. Unlike the *t* test, however, the simple ANOVA can compare the DV across more than two groups. In other words, the IV does not have to be dichotomous. The two-way (two IVs) or multifactor (two or more IVs) ANOVA is used when there is one DV being compared across more than one IV. The multiple analysis of variance (MANOVA) is used when there are more than one DV measured at the interval or ratio level. The one-factor MANOVA is used when there is one IV and more than one DV, and the multifactor MANOVA is used when there are multiple IVs and multiple DVs.

One step further is the analysis of covariance (ANCOVA), also known as the univariate general linear model (GLM). The ANCOVA allows the researcher to examine the differences in one DV across two or more categorical IVs, while controlling for the effects of one or more extraneous variables on the DV. These extraneous variables are called **covariates.** The ANCOVA identifies the difference between the groups on the DV after adjusting for the effects of the covariates or control variables. Chapter 6 will look more closely at this family of statistical procedures.

CORRELATION

While the chi-square test of association examined earlier looks at whether two variables measured at the nominal or ordinal-level are associated with one another, correlation analyses are often used to examine the association between variables measured at the interval or ratio level. Correlation analyses can determine the strength of the relationship between two variables, the direction of the relationship, and the probability that the relationship between the two variables might be due to sampling error. The strength of the relationship is determined by looking at the extent to which the values of the two variables co-vary. If we determine that they do co-vary, the direction signifies whether the values of one variable tend to move in the same direction as the values of the second variable, or whether they tend to move in opposite directions. We can find the *p* value to determine the chance that this relationship is strong enough that we can assume it cannot be explained by sampling error.

Correlation analyses are used for a variety of purposes in research. In descriptive research, correlation analyses can determine the strength of the relationship between two variables of interest with regard to the sample or population being studied. For example, we may want to find out if the number of hours students work while they are completing their college education is significantly related to their overall grade point average in school. Or, we might want to verify that a higher rate of absenteeism from school is significantly related to overall grade point average.

In cases where the correlation between two variables has been shown to be very strong, the value of one variable can be used to predict the value of the second variable. Suppose, for example, that we determine that number of drug arrests in a community has been shown to strongly correlate to the age that an adolescent living in that community will first experiment with illicit drugs. The correlation is a *negative* one, with the higher the number of drug arrests being associated with a younger age of drug experimentation. Social workers could then predict that adolescents living in neighborhoods with high drug arrests may be at high risk for drug use, and could target these neighborhoods for appropriate prevention programs.

Correlation analyses are also used extensively in the development of measurement instruments. They can be used to identify individual questions in an instrument that are weakly correlated to the instrument as a whole, and may, therefore, be deleted from the instrument to reduce its length. Correlation is also used to calculate the reliability of measurement instruments, including inter-rater reliability, test–retest reliability, parallel-forms reliability, and split-half reliability.

Displaying Correlations through Scattergrams

Bivariate correlation analyses are most often used to analyze the **linear correlation** between two variables. Bivariate correlation shows whether the values of the two variables tend to co-vary in the same direction or in opposite directions across all of the values in the sample. In other words, do the values of the first variable tend to increase as the values of the second variable increase or do the values of the first variable tend to increase as the values of the second variable decrease? The extent to which these values co-vary consistently across the entire population or sample determines the strength of the linear correlation. Look again at the example scattergrams shown in Figures 4.1, 4.2, and 4.3. Figure 4.1 suggests that higher levels of depression are associated with higher levels of disordered eating behaviors, and lower levels of depression are associated with lower levels of disordered eating behaviors. This suggests a **positive linear correlation**, as the values of the variables are moving in the same direction. Note in the scattergram, when a line is drawn through the center of the values, it angles from the lower left to the upper right. Figure 4.2 suggests that higher levels of depression are associated with lower levels of disordered eating behaviors, and lower levels of depression are associated with higher levels of disordered eating behaviors. This suggests a **negative linear correlation**, since the values now move in opposite directions. On the scattergram, when a line is drawn through the center of the values, it angles from the upper left to the lower right.

Finally, Figure 4.3 suggests little or no correlation between levels of depression and disordered eating behaviors. The values of the two variables do not co-vary in either direction. The line drawn through the center of the values tends to be a straight horizontal line.

A **perfect positive correlation** is when the values of the first variable move not only in the same direction, but in the exact same amount. A **perfect negative correlation** is when the values of the two variables move in opposite directions, but in the exact same amount. A scattergram displaying a perfect correlation would therefore be a completely straight line. Obviously, variables that have perfect correlations would be rare, and because they move in exactly the same or exactly opposite ways, they would be of little interest to the researcher. Social work researchers are much more interested in variables that have *nonperfect correlations.*

Correlation Coefficients

While scattergrams provide a way of displaying the relationship between two variables, they can only provide an estimation of the strength of the relationship. In order to calculate the exact strength and statistical significance of the relationship, the **correlation coefficient** must be determined. The correlation coefficient provides a numerical indicator of both the strength and the direction of the relationship between the paired values of two interval or ratio-level variables. Because the correlation coefficient provides information about the strength or magnitude of the relationship between the variables, it is considered to be a measure of effect size.

There are several different formulas that can be used to calculate the correlation coefficient, the most common of which is **Pearson's product moment correlation coefficient**, or **Pearson's *r*.** The values of the Pearson's *r* correlation coefficient, expressed as *r*, can range from −1.0, signifying a perfect negative linear correlation, to +1.0, signifying a perfect positive linear correlation. A correlation coefficient of 0 using Pearson's *r* indicates no linear correlation or relationship between the variables at all. Note that the plus sign is usually not displayed before a positive correlation coefficient. The positive nature of the correlation is implied. We want to point out that Pearson's *r* only considers linear correlations between variables. If two variables are related in a nonlinear fashion, another type of analysis would need to be used.

The closer the value of the coefficient is to +1.0, the stronger the positive correlation between the two variables. The closer the value is to −1.0, the stronger the negative correlation. The closer the value is to 0, the weaker the correlation between the variables. It is important to understand that negative correlations are not inherently weaker than positive correlations. It simply indicates that the values co-vary in opposite directions rather than in the same direction. For example, a correlation of .42 and a correlation of −.42 indicate correlations of exactly the same strength. A correlation of −.64 is stronger than a correlation of .39.

Cohen (1988) has developed a system for interpreting the strength of a linear correlation based on the Pearson's *r* correlation coefficient.

He suggests that a correlation coefficient greater than or equal to .1 but less than .3 is a weak correlation, a correlation greater than or equal to .3 but less than .5 is a moderate correlation, and a correlation greater than or equal to .5 is a strong correlation. If your correlation coefficient is negative, simply use the absolute value of the correlation coefficient to interpret the strength of the correlation using Cohen's system.

It is important to point out that even a very strong correlation between two variables ***does not*** indicate that the values of one of the variables actually *cause* the values of the second variable. A third variable could be responsible for causing the values of both of these variables, hence the common phrase, "correlation does not equal causation." If, however, it is determined that one variable has a causal relationship with another variable, these variables must also be correlated with one another.

Determining statistical significance. In addition to looking at the strength and direction of a linear correlation, we can also determine if the strength of the correlation between two variables can be explained by sampling error. Using the correlation coefficient and the sample size, we can identify the *p* value to determine if the correlation between the two variables is statistically significant. Sample size is an important factor when determining the statistical significance of a correlation. A strong correlation of, say, .62, might not be statistically significant with a small sample size, because sampling error may be able to explain the relationship. On the other hand, a weak correlation, for example .25, might be statistically significant with a large sample size, because although the correlation is a weak one, the relationship between the variables could not be explained by sampling error. Therefore, it is important to look at the strength and direction of a correlation, in addition to its statistical significance, when interpreting the outcome of a correlation analysis.

Example. Suppose we want to calculate the strength and direction of the correlation between the number of previous arrests and the number of months to which a group of adolescent males have been sentenced to serve in the juvenile detention center. In this example, our one-tailed hypothesis could read:

Table 4.9 Computing a Pearson's *r* Correlation Coefficient

Client	*Previous Arrests X*	*Months Sentenced Y*	X^2	Y^2	XY
1	0	6	0	36	0
2	1	6	1	36	6
3	2	8	4	64	16
4	1	8	1	64	8
5	3	12	9	144	36
6	2	12	4	144	24
7	5	24	25	576	120
8	1	3	1	9	3
9	3	12	9	144	36
10	1	0	1	0	0
11	2	6	4	36	12
12	4	12	16	144	48
13	0	0	0	0	0
14	1	3	1	9	3
15	3	12	9	144	36
	$\Sigma X=29$	$\Sigma Y=124$	$\Sigma X^2=85$	$\Sigma Y^2=1{,}550$	$\Sigma XY=348$

The number of previous arrests and the number of months to which a group of adolescent males have been sentenced to serve in the juvenile detention center is positively correlated.

The first step is to create columns for the first variable, *X*, the second variable, *Y*, followed by X^2, Y^2, and *XY*. Calculate these values, and then sum each of the columns (see Table 4.9 for computation details).

The next step is to place these values into the Pearson's *r* formula:

$$r=\frac{n\Sigma XY-(\Sigma X)(\Sigma Y)}{\sqrt{[n\Sigma X^2-(\Sigma X)^2][n\Sigma Y^2-(\Sigma Y)^2]}}$$

Replacing the variables for the values in our example, we have:

$$r=\frac{15(348)-29(124)}{\sqrt{[15(82)-29^2][15(1{,}550)-124^2]}}\frac{5{,}220-3{,}596}{\sqrt{(1{,}230-841)(23{,}250-15{,}376)}}$$

$$r = \frac{15(348)-29(124)}{\sqrt{[15(85)-29^2][15(1,550)-124^2]}} = \frac{5,220-3,596}{\sqrt{(1,275-841)(23,250-15,376)}}$$

$$= \frac{1,624}{\sqrt{433(7,874)}} = \frac{1,624}{\sqrt{3,417,316}} = \frac{1,624}{1,848.6} = .879$$

Therefore, r is equal to .879, clearly a very strong positive correlation. To calculate the p value, you must turn to a table that displays the critical values of r (go to the website http://www.radford.edu/~jaspelme/statsbook/Chapter%20files/Table_of_Critical_Values_for_r.pdf for an example table of critical values for Pearson's r). Follow the first column down until you find the sample size for the sample, 15 in this example. Then follow the numbers across until you find a number than is larger that our r value (.879). This value is larger than the value in the right-hand column; therefore, the p value for a one-tailed hypothesis is less than .0005. Clearly the relationship between these two variables is stronger than a relationship that can be accounted for by sampling error. These findings would be presented as follows:

$$r = .879, p < .0005$$

Coefficient of Determination

The **coefficient of determination**, expressed as r^2, can be determined by squaring the correlation coefficient r. This value indicates the proportion of the variation in one of the variables that is related to (or can potentially be explained by) the values of the second variable. For example, suppose the correlation between the number of previous arrests and the number of months to which juvenile offenders are sentenced is .6, the coefficient of determination would be $.6^2$ or .36. This value indicates that about 36% of variation in the sentencing is related to or can be explained by the number of previous arrests. The remaining 64% of the variation is assumed to be related to other variables or error related to the research design.

Examples of Correlations in the Literature

Correlational analyses are used extensively throughout social science research. Kelly, Warner, Trahan, and Miscavage (2009) examined the correlation between level of depression and level of psychological abuse in a sample of 100 women, and found a significant positive correlation between level of depression and level of total abuse (emotional, verbal, physical, and sexual), as well as subscale scores of emotional and verbal abuse.

In another study, Clark, Skouteris, Wertheim, Paxton, and Milgrom (2009) looked at the relationship between depression and body dissatisfaction during pregnancy and the first 12 months postpartum. The researchers found a positive correlation between depression and body dissatisfaction measures and a negative correlation between a depression scale and subscales measuring attractiveness and feelings of strength and fitness. These correlations were strongest at six weeks and six months postpartum when women experienced the most body dissatisfaction.

Correlation is also commonly used when measuring the validity and reliability of measurement instruments. Balfour et al. (2007) used correlation to develop and validate the *HIV Medication Readiness Scale*. This scale assesses the client's psychological readiness for successfully utilizing and adhering to HIV medications. The researchers used correlational analyses to determine that the 10-item instrument demonstrated high internal consistency, test–retest reliability, and sensitivity to change. Other methods were used to assess content validity, convergent validity, and predictive validity.

Using SPSS to calculate the Pearson's *r*. In this demonstration we explain how to conduct a Pearson's *r* Correlation test in SPSS, version 19. The Pearson's *r* Correlation is used to determine whether two interval or ratio-level variables are significantly related to one another. In this example, we use two ratio-level variables from the NELS data set, BY2XRSTD and BY2XMSTD. BY2XRSTD provides a measure of standardized reading scores and BY2XMSTD provides a measure of standardized math scores.

- Open the SPSS data file entitled, BIVARIATE.SAV.
- Click on "Analyze" in the menu bar at the top of the data view screen.
- A drop down menu will appear. Move your cursor to "Correlate."
- Another drop down menu will appear to the right of the first menu. Click on "Bivariate" to open the Bivariate Correlations dialog box.

- Highlight BY2XRSTD in the box on the left side of the Paired Variables dialogue box. Click the arrow to move it to the Bivariate Correlations box.
- Highlight BY2XMSTD in the box on the left side of the Paired Variables dialogue box. Click the arrow to move it to the Bivariate Correlations box.
- Click on the desired correlation coefficient (i.e., Pearson) in the Correlation Coefficients box.
- Click on the desired test (Two-tailed) in the Test of Significance box.
- Click "OK." SPSS should now calculate the Pearson's r Correlation coefficient.
- NOTE: A copy of this output is available on the companion website. The file is entitled, "Pearson r Correlation test results. spv."

Nonlinear Correlations

In this section, we have focused on linear correlations. As noted previously, Pearson's *r* only considers linear correlations between variables. There are, however, other types of correlation, such as *cyclical correlations* and *curvilinear correlation.* A nonlinear correlation would be found when the values of two variables co-vary, but do not continuously vary in the same direction for the entire sample or population. A scattergram displaying a cyclical correlation would look like a series of waves. The values

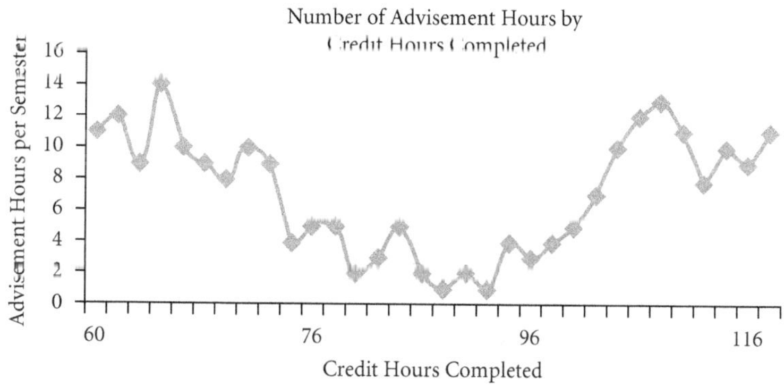

Figure 4.4 Scattergram Showing a Curvilinear Correlation.

of the two variables would alternate moving in the same direction, then moving in opposite directions, etc. A scattergram displaying a curvilinear correlation would form a U shape or an upside-down U shape or arch. For example, the scattergram in Figure 4.4 shows a curvilinear correlation between the amount of time a faculty advisor may spend with their student and the number of credit hours completed by the students.

The graph suggests that when juniors (students with 60 credit hours) first enter a social work program, the faculty advisor spends a significant amount of time with the student. During the second semester of the junior year and the first semester of the senior year (credit hours from 75–105), the faculty advisor tends to spend less time with the student. During the last semester of the program, when students may be completing their internship and are preparing for graduation, the hours spent advising increase. These two variables are clearly correlated; however, it is not a linear correlation.

Alternatives to the Pearson's *r* Correlation

The Pearson's *r* should only be used in cases where the requirements for a parametric statistical test are met. There are alternative statistical tests, such as *Spearman's rho* and *Kendall's tau,* that can be used when the variables are measured at the ordinal-level or when the data are not normally distributed. For example, suppose we want to determine the correlation between socio-economic level and client satisfaction level. Either of these alternative tests could be used to calculate the correlation coefficient. Like the Pearson's *r,* the Spearman's rho and Kendall's tau both have a table of values that must be used to determine whether the correlation is statistically significant.

Bridge to Advanced Statistical Tests

In some situations, it is important to consider the correlation between three or more variables. Multivariate correlation analyses examine the correlation between one outcome variable and several predictor variables. One example is the *multiple correlation (multiple R).* It combines the values of the predictor variables to determine the strength of the correlation between these values and the values of the outcome variable. There are other multivariate analyses that use correlation in their calculations, such as *factor analysis* and *cluster analysis.*

5

Bivariate and Multivariate Linear Regression Analysis

INTRODUCTION

Linear regression analysis is a set of statistical procedures designed to examine relationships between one or more independent variables (IV) and one dependent (i.e., outcome) variable (DV). In linear regression, the DV is always measured at the interval or ratio level. For example, Schiff and Levit (2010) used linear-regression analysis to assess the relationship between the IVs of trauma history and therapeutic alliance between Israeli social workers and female clients receiving methadone treatment and the DV of reduced posttraumatic stress disorder (PTSD). Posttraumatic stress disorder was measured using the Posttraumatic Diagnostic Scale (PDS) (Foa, Cashman, Jaycox, & Perry, 1997). The PDS includes 49 items that assess PTSD symptoms. Responses are measured at the interval level, ranging from 0 to 51.

Bivariate and Multiple Linear Regression Analysis

Linear regression analysis may involve only one IV or it may involve several. Linear regression models with only *one* IV are referred to as **bivariate** or **simple regression models**. Linear regression models with *more than one* IV are referred to as **multiple linear regression models**. In multiple linear regression analysis, each IV provides its own contribution in predicting the DV. Thus, the model being predicted is more complex than with bivariate regression analysis. Multiple linear regression analytic procedures build on bivariate linear-regression analysis. As such, multiple linear regression analysis includes additional requirements and assumptions for running the model, and the interpretation of the results differ somewhat.

Types of Regression Analysis Methods

There are several different types of regression analysis methods. It is important to use the method that matches your research question and the type of data you are analyzing. Linear regression is used when the DV is measured at the interval or ratio-level. For instance, Horwitz-McCue, Balestracci, and Simms (2001) used the linear regression method to examine changes in reported functioning of children in foster care. The DV was assessed using the Vineland Adaptive Behavioral Scale, (Sparrow & Cicchetti, 1987) which is an interval-level measure of foster mother perceptions of foster child adjustment.

Logistic regression is another common regression method. It is used when the DV is measured as a dichotomous variable, with only two values. For instance, studies investigating factors that predict foster care placement—the DV—would apply logistic regression procedures to analyze the data. In this case, foster care placement is a dichotomous variable, with two values—placed into foster care or not. In a real example, Alexander and others (1997) used logistic-regression methods to examine the relationship between individual, family, and school IVs and high school dropout as the DV. High school dropout was assessed as a dichotomous variable with two values—dropped out or not. Linear and logistic regression procedures are only two of several regression methods available for analyzing data. The focus in this book is on linear regression. For more on logistic regression analysis, we recommend Orme and Combs-Orme (2009) or Kleinbaum and Klein (2010).

Structure of the Chapter

In this chapter, we provide a detailed explanation of linear regression analysis, covering the relevant basic statistics (i.e., correlation, properties of a straight line, and variance), sums of squares, model estimation and fit, assumptions and other model requirements, and methods of data entry. We then describe examples of social work studies that have used linear regression analysis methods. At the end of the chapter, we provide step-by-step instructions for conducting linear regression analysis in SPSS, using data from the National Educational Longitudinal Study of 1988 (NELS: 88).

LINEAR REGRESSION ANALYSIS: A DETAILED DESCRIPTION

The goal of linear regression analysis is to determine the best model, in the form of an equation, for predicting values of the DV using the values of the IVs for each case in the dataset. Of course, the model that we develop depends on the data that we have. In this context, the goal is to determine the model that best fits our data. As Tabachnick and Fidell (2007) state, the intent of linear regression is to "arrive at a set of values, or estimated regression coefficients for each of the IVs that bring the *Y* values predicted from the equation as close as possible to the actual *Y* values obtained by measurement" (p. 128).

In linear regression, the best-fitting model is actually represented by a straight line. This straight line connects the **means** of the IVs and the DV in a way that minimizes the sum of the squared distances between the data points and this line. This will be explained in more detail when we discuss the properties of a straight line. In bivariate linear regression, the best-fitting straight line is determined by solving the equation for a straight line, which is $Y' = \ + bX + $, where,

- Y' is the DV or the outcome we want to predict from X. The prime symbol () after "Y" indicates its *predicted* rather than actual value.
- α is the intercept of the line for Y'. It is the value of Y' when X crosses the Y-axis (when $X = 0$).
- b (i.e., beta) is the slope of line for Y'. Its value represents the change in Y for each one unit change in X.

- X is the value of the IV that is used to predict Y'.
- represents the error in the model. Predictions for each case have associated error. The error represents the difference between the scores predicted by the line and the scores that are actually obtained. This will become clearer when we discuss the process for determining the best fitting straight line.

Multiple linear regression analysis includes more than one IV in the model. The equation is changed to reflect the additional IVs, as follows: $Y = \ + b_1X_1 + b_2X_2 + \ldots + b_KX_K + \ $. In this case, b_1 is the value of the estimated coefficient for the first IV (X_1), b_2 is the value of the estimated coefficient for the second IV (X_2), through K IVs. Regardless of whether your model includes one or more than one IV, in linear regression the DV is *always* measured at the interval or ratio level. IVs can be measured at any level (i.e., nominal, ordinal, interval, and ratio), although it is necessary to recode nominal or ordinal level variables with more than two values prior to conducting the analysis. We will discuss this in more detail later in the chapter.

Relevant Basic Statistics in Linear Regression Analysis

Correlation. Linear regression analysis is actually a form of correlation analysis. Recall that correlation analysis produces the Pearson product-moment correlation coefficient, r. It is used to assess the magnitude and direction of the relationship between two variables measured at the interval or ratio level. Linear regression analysis builds on this, going beyond simply assessing relationships to actually predicting the value of the DV from the values of one or more IVs.

The correlation between one IV and one DV in bivariate-regression analysis is straightforward. On the other hand, in multiple linear regression analysis there are three types of correlation—multiple correlation, partial correlation, and semipartial correlation. This is due to having more than one IV in the model, and thus needing to account for how they correlate with one another, in addition to their correlation with the DV. **Multiple correlation** is the linear correlation between all of the IVs in the model and the DV. In the context of multiple linear regression, **partial correlation** is the unique correlation between *one* IV and the DV that is not shared with any of the other IVs. It represents the extent to which an IV and the DV are associated with one another, while controlling for the effects of *all* other IVs on both the DV and the IV. As Tabachnick and Fidell (1996) state, "the contribution of the other IVs is taken out of both the IV and the DV" (p. 157). **Semipartial correlation** is similar to partial correlation in that it is the unique correlation between an IV and the DV. However, in this case, the effects of other IVs are controlled for in either the DV or the IV, but not both. We will review these correlations in more detail later in the chapter when we cover estimation procedures.

The straight line in linear regression. The estimation process in linear regression analysis, referred to as least-squares estimation, involves finding the straight line that best fits the model to our data. Thus, it is useful to describe the properties of a straight line. A straight line is defined by two features—its slope and its intercept. The equation for a straight line is: $Y = \ + bX$, where is its intercept and b is its slope. Kleinbaum and others (1988, p. 44) present a graph of two straight lines with their respective equations as a way to explain the properties of a straight line, replicated in Figure 5.1. To create these lines, arbitrary numbers were selected as values of and b and then plugged into the straight line equations. For line $y = 4 + (-2x)$, $= 4$ and $b = 2$. This describes a straight line with an intercept of 4 and a slope of – 2. For the line $y = 3 + 1x$, $= 3$ and $b = 1$ with an intercept of 3 and a slope of 1.

In the straight line equation, the **intercept**, denoted by , is the point at which the line crosses the vertical (i.e., *Y*) axis of the graph. It is the value of *Y* when *X* equals zero (Keith, 2006). In Figure 5.1, we see that one line crosses the *Y* axis at 4, when $y = 4$ and $x = 0$. We also see that the other line crosses the *Y* axis at 3, when $y = 3$ and $x = 0$. Note that this definition of the intercept is only true when $x = 0$. In multiple regression analysis, the intercept is the mean value of the DV when all of the IVs equal zero (Keith, 2006).

A line's **slope** is its gradient or steepness, denoted by *b*. It represents the number of unit changes in *Y* for a one unit change in *X*. As Kleinbuam and others (1988) explain, because the line is straight, this rate of change is always constant. Thus, for the line $y = 4 \quad 2x$, when x changes one unit, y changes by minus two units (the value of *b*, the slope of the line). For instance, when *X* changes one unit from 3 to 4, *Y* changes minus two units, from 2 [4 – 2(3)] to 4 [4 2(4)], which is minus two units [(2) – (4) = (2) + 4 = 2]. When the value of *X* changes one unit from 1 to 2, the

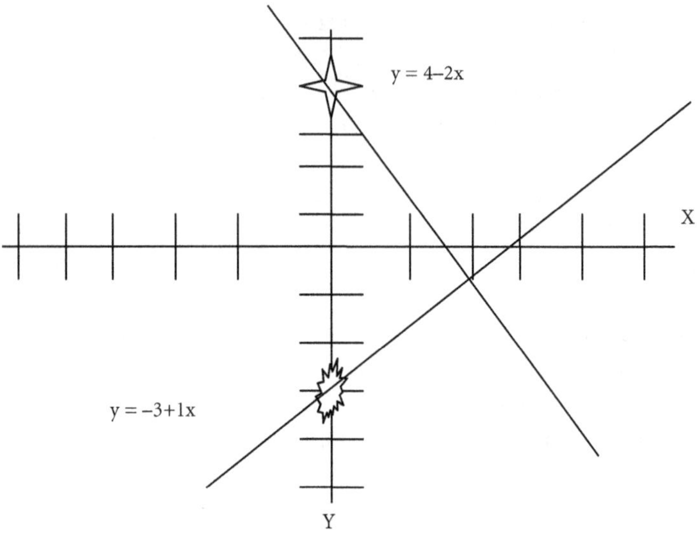

Figure 5.1 Plot of two straight lines. Adapted from Applied Regression Analysis and other Multivariate Methods, by D. G. Kleinbaum, L. L. Kupper, & K. E. Muller, 1988, p. 44, Wadsworth Publishing Company. Copyright 1988 by Wadsworth Publishing Company.

value of *Y* changes minus two units from 2 [4 – 2(1)] to 0 [4 2(2)]. The rate of change is constant.

The sign of a line's slope (either positive or negative) provides information about the nature of the relationship between *X* (the IV) and *Y* (the DV). A line that has a slope with a positive value indicates a positive relationship between *X* and *Y*. In other words, as the value of *X* increases, so does the value of *Y*. A line with a negative slope indicates a negative relationship, such that *Y* decreases as *X* increases. Figure 5.1 shows that, for the line $y = 4 - 2x$, the value of *Y* decreases as the value of *X* increases because the slope is a negative value. The slope of the other line, $y = 3 + 1x$, is positive and as displayed on the graph, as *X* increases, *Y* also increases.

Determining the best-fitting straight line. We will now consider how to determine the best-fitting straight line to solve the regression equation (which is the same equation for a straight line). In linear regression the **least-squares-estimation method** is used to find the best-fitting straight line. This method is used to calculate the slope and intercept as

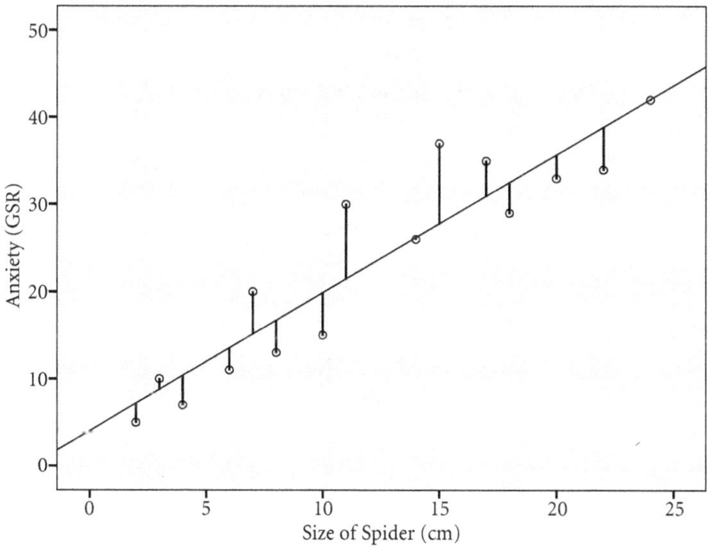

Figure 5.2 Scatter plot of several data points with the best fitting straight line to show vertical differences between the line and some of the data points as an illustration of residuals. Adapted from Discovering Statistics using SPSS, by A. Field, 2005, p. 147, Sage Publications. Copyright 2005 by Sage Publications.

a representation of a line that (a) provides the best fit with the data, and (b) minimizes the sum of squared differences (i.e., residuals) between the predicted data points on the line and the actual observed data points. The differences between the predicted and actual data points represent the error between what we predicted and what we actually obtained. These differences are referred to as **residuals.**

Figure 5.2 illustrates what we mean by best-fitting line and residuals (Field, 2005, p. 147). The figure provides a scatterplot of observed or actual data points and a line drawn within this set of points. The data points represent the *X* and *Y* values for each case. In this example, the *X* values represent the size of a spider, and the *Y* values represent anxiety. The overall pattern of data points suggests that, as spider size increases, so too does anxiety. The line on the graph is what we use to predict the values of *Y* from the values of *X*. It is the line that fits the data most closely. What does this mean? We can see that two points fall exactly on the line, and the rest fall above or below it. In terms of residuals, our interest is in the points that fall off the line, in particular their vertical distance from the line. This vertical distance is the difference between the model fitted to the data (the straight line) and the actual data. Each set of points that falls off the line has a vertical-difference value. This value represents the residual or error from the straight line for that set of points.

The least-squares method of estimating the best-fitting straight line is determined by these residuals (Field, 2005). The values of the residuals (i.e., their differences) indicate how well the line fits the data. The residual values of the points that fall above the line are positive. The residual values of the points that fall below the line are negative. If we simply added the negative and positive differences together, they would likely cancel each other out. We can square these differences prior to summing them, to avoid this. This is referred to as the *sum of squared differences*. The sum of squared differences is used to assess the accuracy of our model. The smaller the value of the sum of squared differences, the more closely the line represents the data. The least-squares method involves determining the line that has the lowest sum of squared differences. This is the line that best represents the actual data.

Sums of Squared Differences

The least-squares-regression method is used to predict the best-fitting regression line to our data. But how well does this line fit the data? It may be the best-fitting line, but the fit may not be very good. In order to determine the line's quality of fit (i.e., **goodness of fit**), we compare it to the basic model (i.e., the mean value of *Y*). This indicates the extent to which the predicted line improves the fit, relative to the basic model. The predicted line should provide a better fit to the data, compared to the basic model. To determine the goodness of fit, we compare differences between the observed data points and the mean value of *Y*, to the differences between observed data points and the predicted regression line. The basis of this comparison is made from three types of sums of squared differences—total sum of squared differences (SS_T), residual sum of squared differences (SS_R), and model sum of squared differences (SS_M). By convention, these sums of squared differences are typically referred to as total sum of squares, residual sum of squares, and model sum of squares. Field (2005, p. 149) presents a set of three graphs that illustrates how these sums of squares are determined. These graphs, provided in Figure 5.3, are described as follows.

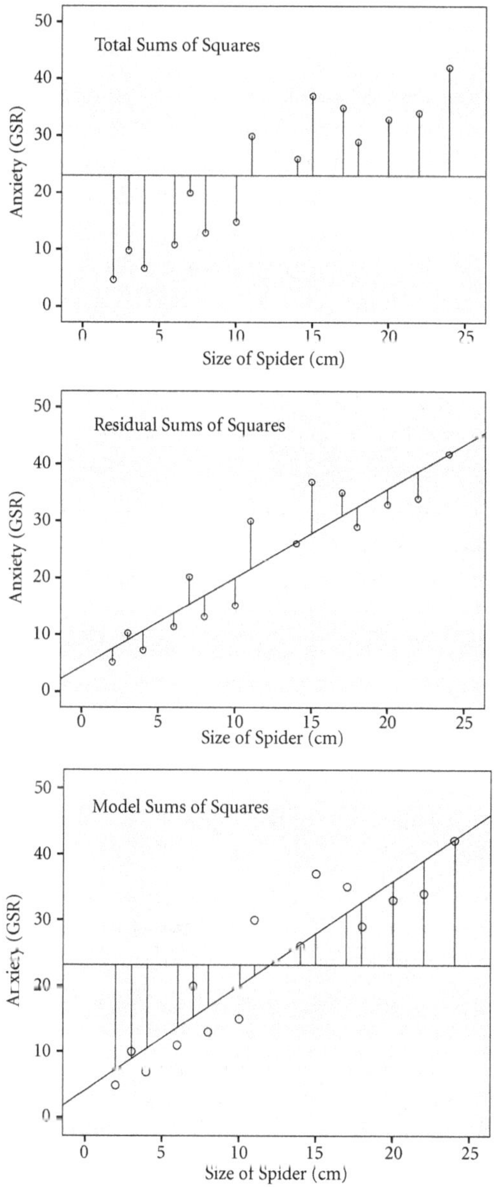

Figure 5.3 Three scatter plots showing how total sums of squares, residual sums of squares, and model sums of squares are derived. Adapted from Discovering Statistics using SPSS, by A. Field, 2005, p. 149, Sage Publications. Copyright 2005 by Sage Publications.

The basic model represents the **total sum of squares** (SS_T). It is the difference between the observed data and the mean value of *Y* across all observations or cases. The formula for determining $SS_T = (Y \ \bar{Y})^2$. The value of SS_T is an indicator of how well the mean of *Y* represents the actual data. The graph at the top of Figure 5.3 provides an example. As the diagram shows, the mean value of *Y* is 23, indicated by the straight horizontal line in the middle of the graph. The data points for the observed data represent all cases that fall above or below the line (i.e., the mean value of *Y*). To get the value of SS_T, the distances between each data point and the mean value of *Y* are squared and then summed.

The total sum of squares is portioned into two components—sum of squares predicted by the regression equation, called **model sum of squares** (SS_M), and the left over sum of squares, called **residual sum of squares** (SS_R). This relationship is represented as $SS_T = SS_M + SS_R$. The residual sum of squares is the difference between the predicted data points on the line and the actual observed data points. The residual sum of squares represents the errors in the predictions. The formula for $SS_R = (Y \ Y)^2$, where *Y* represents the actual scores and *Y* represents the predicted scores. SS_R is the error in what was predicted. The second graph in the middle of Figure 5.3 provides an example of the differences between the actual data points (*Y*) and the predicted regression line (*Y*). As the graph indicates, only two data points fall on the predicted regression line. Eight data points fall below the line and five data points fall above the line, at varying distances. The squared sum of these distances is the error in the prediction.

The sum of squares predicted by the regression equation is called the model sum of squares (SS_M). The model sum of squares represents the extent to which the prediction improves the fit of the model to the data, relative to just using the mean of *Y*. The formula for $SS_M = (Y \ \bar{Y})^2$ where *Y* represents the predicted scores. If we know the values of SS_R (residual sum of squares) and SS_T (total sum of squares), we can simply subtract SS_R from SS_T to get the value of SS_M. Tabachnick and Fidell (1996) describe SS_M as the additional proportion of variation in *Y* that is explained by adding IVs to the basic model. The bottom graph in Figure 5.3 provides a plot of the differences between the predicted regression line (*Y*) and the mean value of *Y* (the basic model). As the graph indicates, the predicted line provides a better fit to the data, relative to the basic model (SS_T). That is, the distances between the actual data points and the line representing the basic model are greater than the distances between the actual data points and the predicted regression line.

A comparison of the magnitudes of SS_M and SS_T provides an indication of the goodness of fit of the model to the data. If the value of SS_M is large, then the regression model (SS_M) is better than simply using the mean (SS_T) to predict the DV. This implies that the regression model has improved upon how well the outcome variable can be predicted. However, if the value of SS_M is small, then this suggests little improvement in using the regression model to predict the DV, relative to the mean.

Estimating the Model and Evaluating Fit

Proportion of variance explained. The values of SS_M and SS_T are used to calculate a statistic, R^2, as an indicator of goodness of fit, or the proportion of improvement due to the model. It is a measure of the magnitude of the effects of the IVs in explaining variation in the DV, and thus is considered to be an effect size measure. R^2 is also referred to as the **squared multiple correlation coefficient** or the **coefficient of multiple determination.** It is calculated by dividing the model sum of squares, SS_M, by the total sum of squares, SS_T, or $R^2 = SS_M \;\; SS_T$. If SS_T represents the total amount of variance in the DV to be explained, and SS_M is the amount of variance in the DV explained by the predicted model (SS_M), then R^2 indicates the proportion of variance in the DV that is explained by the model. Owing to sampling fluctuations and the fact that R^2 can never be negative, it tends to be overestimated, particularly in smaller samples. Statistical software packages, including SPSS, provide an adjusted R^2 to account for this overestimation. Both the original and adjusted values of R^2 are provided in SPSS output. A confidence interval (CI) can be calculated to assess the relative closeness of R^2 to its population value.

Let's examine in more detail exactly what R^2 measures. In our previous discussion on correlation we introduced the terms multiple correlation, partial correlation, and semipartial correlation. The statistic R^2 indicates multiple correlation, squared. The statistic R (i.e., multiple correlation coefficient) indicates the correlation between the predicted and actual values of the DV. In other words, R provides a measure of the degree to which the set of IVs together correlates with the DV. In multiple regression analysis we wish to know the proportion of variance in the DV that is explained by the IVs. The multiple correlation coefficient, squared (i.e., the coefficient of multiple determination, or R^2) provides a measure of this.

In determining the correlation between a set of IVs and a DV, it is necessary to account for the potential correlation between the IVs themselves. Otherwise, the total correlation will be greater than 100%. For example, consider the study by Nasim, Belgrave, Corona, and Townsend (2009) designed to determine factors that predict alcohol-refusal attitudes among teenagers. The DV is alcohol-refusal attitudes and the IVs are lifetime alcohol use, number of alcohol-using peers, and number of tobacco-using peers. For the purpose of explaining why it is necessary to account for overlapping correlation between IVs, we assume that bivariate correlation analyses between each IV and the DV indicate the following correlation coefficients: $r = .60$ for lifetime alcohol use; $r = .80$ for number of alcohol-using peers, and $r = .55$ for number of tobacco-using peers. To determine the proportion of variance in the DV that is explained by these IVs, we calculate the coefficient of multiple determination (R^2), which is simply the value of R, squared. In this case, the values of R^2 for each of the IVs are: .36, .64, and .30. The sum of these values is 1.30, which exceeds 100%. This is an example of what happens when we fail to account for the overlapping correlations that the IVs have with one another.

To account for overlapping correlation, we must calculate either partial or semipartial correlation to determine the unique correlation between the IV and the DV (Field, 2005). **Partial correlation** (i.e., partial r) is the unique correlation between two variables, separate from the other variables. Given three variables, X_1, X_2, and X_3, the partial correlation between X_1 and X_2 is estimated such that the extent to which both X_1 and X_2 are correlated with X_3 has been partialled out or removed. **Semipartial correlation** is the unique correlation between two variables, such that the influence of a third variable has been removed from only *one* of the other two variables. Given three variables, X_1, X_2, and X_3, the semipartial correlation between X_1 and X_2 is estimated such that the extent to which *either* X_1 *or* X_2 is correlated with X_3 has been partialled out or removed. Tabachnick and Fidell (1996) indicate that semipartial correlation is more useful in determining the relative importance of an IV, compared to partial correlation. This is because the contribution of other IVs (e.g., X_2 and X_3) to the DV is removed only from X_1. When the semipartial correlation statistic is squared, it provides an indicator of the unique contribution of X_1 to the total variance of the DV.

The F test and evaluating overall model fit. The R^2 statistic indicates the proportion of variance explained by the IVs in the model. We use the *F*-test to evaluate whether R^2 is statistically significant. It tests the null hypothesis that none of the IVs are significantly different from 0. The *F*-test assesses improvement in the model (SS_M), compared to model error (SS_R). The sums of squares are affected by the actual number of data points that are summed. Thus, in order to reduce this bias, the *F*-test is based on the **mean sum of squares** for both the model and the residuals. The mean sum of squares for SS_M and SS_R (i.e., MS_M and MS_R, respectively) are calculated by dividing each by its degrees of freedom. The degrees of freedom for the SS_M are the number of IVs in the model; the degrees of freedom for the SS_R are the number of estimated coefficients including the intercept subtracted from the number of cases. The formula for the *F* test is: $F = MS_M \; MS_R$. We evaluate the *F*-test by comparing the results of the ratio to a critical value of *F*, based on the degrees of freedom in the numerator and the denominator, and the *p* value. If MS_M is large, relative to MS_R, then the corresponding *F* ratio will be large. This is also likely to indicate that our model provides an improvement in predicting *Y*, relative to the residuals.

The t test and evaluating estimated coefficients. We also evaluate the statistical significance of the individual predictors, or the *b* coefficients. In this case, we are interested in knowing whether the *b* coefficients are statistically different from 0. As discussed previously, each *b* represents the slope of its corresponding IV. Its value shows the change in the DV as a result of a unit change in the IV, as a measure of the effect of the IV on the DV. The *t* test is used to assess the null hypothesis, H_0: $b_i = 0$. The closer the slope is to 0, the *less likely* that a one unit change in the IV will result in any change in the DV. A slope that is at or close to 0 also means that the regression line is flat, as illustrated in the graph at the top of Figure 5.3, which represents the basic model using the mean value of *Y*. One can calculate a CI for each *b* coefficient to assess how close the estimate is to its true population value.

Assumptions and Other Model Requirements

Linear regression analysis is a parametric test that is used to make inferences about a population. Thus, the four assumptions described in Chapters 1 and 4—DV measured at the interval or ratio-level data, normally distributed data, homogeneity of variance, and independence of observations—apply to this procedure. In addition, because linear regression analysis is a form of the general linear model, other assumptions and model requirements are necessary, and are described as follows:

- **The relationship between the IVs and the DV is linear.**

The idea here is that Y_i is expected to increase by b_{YX} units for every one unit increase in X_i, because the relationship between X_i and Y_i is assumed to be linear. One common method for evaluating the linearity assumption is to produce a scatterplot of the pairs of scores for the DV against each IV and then examine the shape of the distribution to assess linearity. The companion website of this book includes step-by-step instructions for assessing the linearity assumption using a scatterplot. Another common method is to plot the standardized residuals against the standardized predicted values (Lomax, 1992). In either case, if nonlinearity is present, the overall shape of the distribution will be non-linear. Minor deviations do not have a major impact on regression results. For serious violations, the data can be transformed in a way that increases linearity. Field (2005) provides a thorough discussion of transforming nonlinear data.

If the relationship is not linear, then the amount that Y_i is expected to increase depends on the particular value of X_i (Lomax, 1992). Using a variation of an example by Keith (2006, p. 170), consider the level of anxiety in passing a standardized preliminary examination as the IV, measured on a three-point scale (i.e., low, average, and high), and actual performance on the prelim exam as the DV, measured as a ratio-level variable on a scale from 0 to 100. Imagine that on average, students who have little anxiety about the exam score low ($M = 20$) and students who have high anxiety also score low ($M = 20$). Imagine further that students who have an average amount of anxiety score much higher ($M = 80$). If we were to graph these means, we would observe that prelim test anxiety has a curvilinear relationship with prelim test results such that test scores vary based on level of anxiety, with low and high anxiety associated with low scores and average anxiety associated with higher scores. The implications for analysis with curvilinear relationships are beyond the scope of this book. If you confront this situation in your analysis, we suggest that you consult Keith as he provides detailed instructions for dealing with curvilinear relationships.

- **The errors are independent of (i.e., uncorrelated with) one another. This is referred to as a lack of autocorrelation.**

This assumption means that the distribution of the error values follows a random rather than systematic pattern. Violations of this assumption occur when errors are positively or negatively correlated with one another. Violations affect the accuracy of the standard error estimates, which can increase the likelihood of making a Type I or Type II error. Violations can be detected by examining the shape of the residual plot (e.g., e_i vs. X_i) to determine the extent of randomness in the distribution, or by examining the results of the Durbin–Watson test. The Durbin–Watson test is available in SPSS and other statistical software programs.

Some research designs are more vulnerable to violations of the independence of errors assumption than others. Longitudinal designs, designs with data collected from multiple groups of people in different locations (e.g., children in classrooms), and designs that involve replication are more likely to result in violating the independence of errors assumption, relative to other designs. On the other hand, this assumption is usually met when participants are randomly assigned to groups or with designs that include only one observation (Lomax, 1992).

- **The variance of the errors should be constant for all values of *X*.**

This is referred to as **homogeneity of variance** or **homoscedasticity.** It means that the variances should be equal for all values of the IVs. A violation of this assumption (i.e., unequal variances) is referred to as **heteroscedasticity.** Violations do not necessarily affect estimated regression coefficients. However, violations may result in inflated standard errors, and thus lead to an increased likelihood of making a Type II error (failing to reject a false null hypothesis and discarding an IV that is actually significant).

SPSS offers three ways to determine whether the homogeneity of variance assumption has been violated. First, one can produce a plot of the standardized residuals against the standardized predicted values and then examine the shape of the distribution. Data points that are evenly dispersed indicate that the assumption of homoscedasticity has been met. Another method is to examine the results of the Levene test for homogeneity of variance. If the test statistic is nonsignificant, then we retain the null hypothesis that the difference between the variances equals 0, and the assumption has been met. A third method is to calculate and examine the variance ratio. SPSS offers the Hartley's F_{MAX} procedure as a way to do this. Note that the Levene's test is affected by sample size such that, in large samples, its results can be significant even when variances are not different. We recommend examining the plot of the standardized residuals and the Hartley's F_{MAX} to confirm results of the Levene's test.

What can be done to correct for violations of this assumption? Minor violations have a negligible impact on the results. To address serious violations, however, one can consider what is referred to as variance-stabilizing transformations, such as taking the log or square root of Y (Lomax, 1992). Another option is to use a different method of estimation that does not require homogeneity of variance, such as the weighted-least-squares method.

- **The distribution of the residuals should approximate the shape of a normal curve.**

Errors are assumed to be normally distributed, roughly in the shape of the normal curve. A nonnormal distribution of the residuals affects the precision of the estimated coefficients. The risk of violating this assumption is higher when observations include outliers, when samples are small (< 25 per group), and when sample sizes across groups are not equal (Lomax, 1992). Several graphing options are available to assess the extent to which the shape of the distribution of errors is normal (e.g., histograms, stem and leaf plots, box plots, or Q-Q plots). In general, even with moderate violations of the normality assumption, results are fairly robust. In severe cases, one can apply techniques such as logarithmic transformation to modify the shape of the distribution, or use tests that do not require normal distributions of errors (i.e., nonparametric tests).

Other Model Requirements

- **The model should be correctly specified.**

For a model to be correctly specified, several conditions must be met. First, all relevant IVs must be included in the model. Second, all irrelevant variables should be excluded from the model. Third, each IV should be measured without error. Finally, the model's IVs should not be correlated with variables that are not in the model. If an IV is correlated with a variable that has not been included in the model, the coefficient estimate for the IV in our model may not accurately reflect the true relationship between the IV and the DV. In other words, this relationship may be confounded by a third variable outside of the model.

- **The independent variables in the model should not be strongly correlated with one another (i.e., no multicollinearity).**

When IVs are strongly correlated with one another, this is referred to as **multicollinearity**. IVs that are strongly correlated with one another are basically measuring the same thing. Thus, the values of the estimated coefficients for strongly correlated IVs would be interchangeable, making it nearly impossible to determine their unique contribution to explaining the variance in the DV.

Multicollinearity leads to increased chances of underestimating the multiple correlation coefficient because the *unique* contribution of highly correlated IVs is not being accounted for. As Field (2005) points out, increased collinearity leads to increased standard errors of the estimated coefficients, which increases the possibility of making a Type II error.

Multicollinearity can be assessed by examining the correlation matrix for excessively high correlations between pairs of variables. Correlation values greater than .80 ($r > .80$) between two IVs may indicate multicollinearity (Abu-Bader, 2006, p. 102). Also, SPSS offers tests to conduct collinearity diagnostics. One of the more common tests is the variance inflation factor (VIF). A VIF value of 10 or greater suggests that multicollinearity between two IVs in the model is likely (Wetherill, 1986). Multicollinearity can be addressed in at least one of two ways—by removing one of the IVs from the model that is causing the problem or by combining the IVs to create a composite variable (Abu-Bader, 2006).

- **There should be no influential outliers among the IVs or in the solution.**

One important requirement in linear regression analysis is the absence of extreme cases, or cases with values that have an unusually large impact on the results, relative to the other cases. These types of cases are referred to as **outliers**. Outliers can affect the results in two ways. First, cases with values that are smaller or larger than the other cases can inflate or deflate correlation coefficients. In other words, outliers can bias the results because they affect the value of the estimated coefficients. Second, outliers can also affect the intercept and/or the slope of the regression line such that the intercept may cross the *Y*-axis at a lower or higher point or the slope can become steeper or flatten. SPSS and other statistical software programs offer several statistics that can be used to assess the impact of outliers on the results. We will go over these statistics and the process of determining the impact of outliers in the example at the end of the chapter.

Outliers that have extreme residual values can also affect regression results. Any case with a large residual, relative to the other cases, may be an outlier. At least two methods are available to identify residuals with extreme values. Scatterplots of residuals provide a visual display that can be used to identify cases with extreme residuals. We can also convert the residuals into a standard unit of measurement to produce standardized residuals, and then determine whether any have an absolute value greater than 3. If so, this is indicative of a case that has an unusually large influence on the results (Tabachnick & Fidell, 1996).

- **The sample size should be large enough to detect results at the desired effect size.**

The minimum acceptable sample size (i.e., number of cases) for linear regression analysis depends on the size of the effect we hope to detect (i.e., small, medium, or large), the alpha level selected for statistical significance (e.g., .05), and the level of power it will take to detect the effect. A number of computer programs are available to calculate minimum sample size based on desired effect size. Two examples are Sample Power, which is available as part of SPSS Add On, and G*Power, which can be downloaded from the Internet at no cost.

- **The IVs can be measured at any level of measurement (i.e., nominal, ordinal, interval, or ratio).**

While IVs can be measured at any level of measurement in multiple regression analysis, it is necessary to modify (i.e., recode) nominal or ordinal-level IVs that have more than two values or categories (e.g., race), prior to conducting the analysis. This is because linear regression procedures can only handle interval or ratio-level IVs, and nominal or ordinal-level IVs with two values only (i.e., dichotomous). Thus, if a nominal or ordinal-level IV includes more than two values, it is necessary to recode that variable into a series of categorical IVs with only two values each. For example, we may have an IV that indicates a respondent's marital status with four categories—married, unmarried, divorced, or widowed. In order to measure whether a respondent's particular marital status is related to the DV, we must recode the marital status variable into four new variables, each with two values—married or not, unmarried or not, divorced or not, and widowed or not. This recoding procedure is called dummy coding. The companion website for this chapter provides an example of dummy coding with step by step instructions for conducting this procedure in SPSS.

An Additional IV Consideration: Interactions

Interaction effects occur when the impact of one IV on the DV depends on the value of another IV. Thus, the impact on the DV is the combined effect of the two IVs. Interaction effects can be examined in multiple linear regression analysis. To test for interaction effects of two IVs on the DV in multiple regression models, it is necessary to create an IV that represents the interaction effect. This is done by multiplying together the two IVs in question. The results are then used to create the interaction variable. Interactions between IVs measured at either the nominal or interval/ratio-levels can be examined; this is one of the advantages of multiple linear regression analysis compared to two-way ANOVA, which can only test for interactions between nominal-level variables. A detailed presentation of testing interactions in multiple regression analysis is beyond the scope of this book. For more information on this topic, see Keith (2006).

Methods of Data Entry in Multiple Linear Regression Analysis

Three methods of data entry are available in conducting multiple linear regression analysis—**simultaneous (i.e., forced entry or standard) multiple regression**, **sequential (i.e., hierarchical) multiple regression**, and **statistical (i.e., stepwise) multiple regression**. These methods are differentiated from one another based on the way in which the IVs are entered into the equation when running the model. What drives the decision to use one method over another is the extent to which theory or previous research leads us to expect that the IVs will be correlated with one another. In other words, what is known about their shared or overlapping variance in explaining the DV?

A very important point here is that the output differs depending on which method is used. This is because of differences in the order of entry of the IVs, and the fact that the order of entry provides different information about the effects of the IVs on the DV. In simultaneous multiple regression, all IVs are entered into the equation at the same time. Simultaneous multiple regression analysis provides information about the direct effects of the IVs on the DV. This method is used when IVs are expected to contribute fairly equally in explaining the DV and when they are not expected to be highly correlated with one another.

In sequential or hierarchical multiple regression, IVs are entered into the equation in a predetermined order in multiple steps, either individually or as groups. This order is determined based on existing knowledge about the level of logical or theoretical importance of particular IVs in predicting the DV, or based on time precedence (Keith, 2006). One focus of sequential multiple regression analysis is the magnitude of the change in R^2 from one step to the next. Sequential multiple regression analysis provides information about the total effects of the IVs on the DV, including indirect effects of IVs. The order of entry of the IVs is critical, as the output may vary depending on the sequence in which IVs are entered in the model. Regardless of the predetermined order of entry, you must be able to defend and justify your decisions.

Finally, statistical multiple regression involves using a statistical procedure rather than theory to determine the order of entry for each IV. As such, the computer selects the order of the IVs rather than on the basis of theory or prior research. This method should be used with caution and primarily for exploratory purposes.

Research Examples using Multiple Linear Regression Analysis

Linear regression analysis is an appropriate method to use as a data-analytic procedure when the intent is to determine factors that predict an outcome (the DV) measured at the interval or ratio-level. Income is frequently measured at the ratio level and is appropriate as a DV in multiple linear regression analysis. Linear regression analysis is also often used in studies that use scales to measure outcomes, such as the Child Behavior Checklist (CBCL), used to rate child problem behaviors, or the Center for Epidemiologic Studies Depression Scale (CES-D), which assesses depression. Shiovitz-Ezra and Leitsch (2010) used the Revised UCLA Loneliness Scale (R-UCLA) (Russell, Peplau, & Cutrona, 1980) to examine the relationship between perceived and actual social ties (the IVs) and loneliness (the DV) among the elderly. The DV was measured at the interval level. The authors found that the number of friends and relatives, the frequency of their contact, and the quality of the interactions were negatively related to loneliness among older adults (i.e., as the values of the IVs increased, the value of the DV decreased).

In another example, Nasim et al. (2009) used sequential linear regression analysis to identify risk and protective factors (the IVs) associated with alcohol and tobacco refusal attitudes (the DVs) among 227 African-American youth across individual, family, peer, and community domains. The authors used the Specific Events Drug and Alcohol Refusal Efficacy (SEDARE) scale (Connors, Plybon, & Pegg, 2003) to measure the DV. The scale provides an assessment of refusal attitudes at the interval-level of measurement. Sets of domain specific IVs were entered into the equation in the following order: (a) individual, (b) peer, (c) family, and (d) community, based on their expected level of importance in predicting the DV. This allowed for an assessment of the relative effects of groups of IVs within and across domains. As expected, findings revealed differences between domains with regard to their relationship to tobacco- or alcohol-refusal attitudes. Individual variables were more strongly related to both alcohol- and tobacco-refusal attitudes, compared to variables in the other domains.

CONDUCTING LINEAR REGRESSION ANALYSIS IN SPSS: AN EXAMPLE

The following provides instructions for conducting multiple linear regression analysis in SPSS, using data from NELS:88. We also provide instructions to check for multicollinearity and for the normality and homogeneity of variance assumptions. A more detailed set of instructions is available on the companion website of this book.

The DV is base year standardized reading scores (BY2XRSTD); the IVs are socioeconomic status (BYSES), family size (BYFAMSIZ), self-concept (BYCNCPT1), urban residence, (URBAN), rural residence (RURAL), and sex (GENDER). Although not used in this example, the data set also includes two identification variables [student ID (STU_ID) and school ID (SCH_ID)] and one weight variable (F2PNWLWT).

- Access the companion website to open the SPSS data file entitled, REGRESSION.SAV.
- Click on "Analyze" in the menu bar at the top of the data view screen.
- A drop down menu will appear. Move your cursor to "Regression."
- Another drop down menu will appear to the right of the first menu. Click on "Linear" to open the Linear Regression dialog box.
- Click to highlight BY2XRSTD in the box on the left side. Click the top arrow to move it to the Dependent box.
- Click to highlight BYSES, BYCNCPT1, and GENDER and move each of them one at a time to the Independent(s) box.
- Click "Statistics" in the upper right corner to open the Linear Regression: Statistics dialogue box. Make sure that "Estimates" and "Model fit" are checked. Check "Collinearity Diagnostics."
- Click "Continue" to return to the Linear Regression dialogue box.
- Click "Plots" just below "Statistics" to open the Linear Regression: Plots dialogue box. Click and move "*ZRESID" to the Y box. Click and move "*ZPRED" to the X box. Check "Histogram."
- Click "Continue" to return to the Linear Regression dialogue box.
- Click "OK" at the bottom of the Linear Regression dialogue box to run the multiple linear regression analysis.
- NOTE: A copy of this output is available on the companion website. The file is entitled, "Multiple Linear Regression results. spv."

ADDITIONAL READING

Cohen, J., Cohen, P., West, S., & Aiken, L. (2002). *Applied multiple regression/correlation analysis for the behavioral sciences* (3rd ed.). London: Routledge Academic.

Field, A. (2009). *Discovering statistics using SPSS* (3rd ed.). Thousand Oaks, CA: Sage Publications.

Keith, T. Z. (2006). *Multiple regression and beyond*. New York: Allyn & Bacon.

Tabachnick, B. G. & Fidell, L. S. (2007). *Using multivariate statistics* (4th ed.). New York: Allyn & Bacon.

6

Analysis of Variance (ANOVA) and Covariance (ANCOVA)

INTRODUCTION

This chapter provides a description of **Analysis of Variance (ANOVA)** and **Analysis of Covariance (ANCOVA)** methods of data analysis. ANOVA and ANCOVA methods are used to compare means across different groups. Recall that the *t* test is also used to compare means between groups. The *t* test, however, is used to compare means between *no more* than two groups, whereas procedures from the ANOVA family can be used to compare means between *more* than two groups. ANOVA procedures are often used in experimental research when the investigator is interested in comparing outcomes across treatment and control groups. For instance, Lindsey, Hayward, and DePanfilis (2010) used ANOVA procedures to assess differences in outcomes between four groups of people randomly assigned to one of the following conditions: (a) enhanced intervention for three months, (b) enhanced intervention for nine months, (c) regular intervention for three months, and (d) regular intervention for nine months, in their evaluation of Family Connections, a ɔram designed to prevent behavioral problems among children at risk ıf neglect.

As mentioned in the Introduction, while ANOVA and multiple regression analysis share the assumption of linearity in the relationship between *X* and *Y*, there are important differences between the two methods (Keith, 2006). For instance, linear regression models include IVs measured at any level whereas ANOVA models examine the impact of IVs measured only at the nominal level. Also, analyses typically involve assessing the impact of several IVs in multiple linear regression; the number of IVs in ANOVA is limited. Finally, ANOVA procedures are often used in intervention research, where the goal is to determine the impact of a treatment on an outcome. Linear regression analysis can be used for this purpose, and it can also be used in nonexperimental research where the intention is to simply describe relationships.

Types of ANOVA Procedures

The ANOVA family includes several types of statistical tests. Commonly used tests are one-way ANOVA, two-way ANOVA, factorial ANOVA, multivariate ANOVA (also referred to as MANOVA), ANCOVA, and repeated measures ANOVA. These tests share two features—the DVs are always measured on an interval or ratio-level and the IVs are always measured at the nominal level. The tests differ with regard to the number of DVs and IVs in the model. The first three tests (i.e., one-way ANOVA, two-way ANOVA, and factorial ANOVA) are based on models with *only one* DV; they differ in the number of IVs in the model. **One-way ANOVA** is a bivariate test used to examine models with one DV and one IV. In **two-way ANOVA**, the model includes one DV and two IVs. Factorial ANOVA tests models with one DV and three or more IVs. MANOVA tests are conducted on models with more than one DV and IV. ANCOVA is another frequently used procedure from the ANOVA family. In addition to one DV and one IV, ANCOVA models include what is referred to as **covariates**. Covariates are explanatory variables that are hypothesized to explain some of the variance on the DV, beyond that of the IV. We will explain "covariates" in more detail later in this chapter.

One-way ANOVA, two-way ANOVA, factorial ANOVA, multivariate ANOVA, and ANCOVA also share the feature that all sample participants must be members of only one condition or level of the IV so that participant scores are independent (i.e., mutually exclusive) across levels. The repeated-measures ANOVA procedure does not have this requirement. Participants can be assessed under more than one condition in the experiment. Repeated-measures ANOVA is often used in longitudinal studies that employ experimental or quasiexperimental designs with multiple data collection follow-ups. For example, in their study to assess the effectiveness of parent-child interaction therapy, Leung, Tsang, and Heung (2009) used repeated measures ANOVA to analyze data collected at three points—pretest, posttest, and six month follow-up. In this chapter, we explain one-way and two-way ANOVA, and ANCOVA.

Structure of the Chapter

This chapter is divided into four sections. In the first section we provide a detailed explanation of one-way ANOVA, including relevant basic statistics that inform ANOVA models (i.e., mean and variance), the ANOVA Source table (i.e., components of the actual analysis), estimation procedures, model assumptions, and procedures for making multiple comparisons. We conclude this section by providing a description of examples from social work research that have used one-way ANOVA methods in analyzing data. In the second section, we explain two-way ANOVA. This presentation begins with a general description and purpose. Then we define important terms and describe model assumptions, hypothesis testing, and partitioning sums of squares. Examples of social work research using two-way ANOVA are provided at the end of this section. The third section provides an overview of ANCOVA, including key concepts, model requirements and assumptions, and social work research examples using ANCOVA methods. In the final section, we provide step-by-step instructions for conducting one-way ANOVA in SPSS, with data from the National Educational Longitudinal Study of 1988 (NELS: 88).

ONE-WAY ANALYSIS OF VARIANCE (ANOVA): A DETAILED DESCRIPTION

ANOVA models are designed to test the null hypothesis that the population means between two or more groups are equal. This is done by evaluating the extent to which observed scores from the sample are similar to the population scores. The notation is: H_0: $\mu_1 = \mu_2 = \mu_3 = \ldots\mu_j$ where μ is the population mean for all j groups in the population. The alternative hypothesis is that not all means are equal (i.e., H_A: not all the μ_j are equal). Because ANOVA is used to assess differences between group means, the DV must be measured in a way that a legitimate mean can be determined. This requires the DV to be measured at the interval or ratio-level. For example, Lindsey and others (2010) used responses to the Child Behavior Checklist (CBCL) as the DV to examine differences in child outcomes. The IV was a categorical variable—assignment to one of four intervention conditions. In ANOVA, the IV is categorical and represents membership in one of multiple groups within which to compare means on the outcome.

The equation for the one-way ANOVA population model is:

$$Y_{ij} = \mu + \alpha_j + \varepsilon_{ij},$$

such that
Y_{ij} = observed score on the DV for individual i in group j,
μ = population overall (i.e., grand) mean,
α_j = group effect for the jth group, and
ε_{ij} = residual error for individual i in group j.

Similar to multiple regression analysis, the **residual error** in ANOVA is unexplained **variance**, or the portion of Y that is not explained by the IV. As in multiple regression analysis, residual error may be due to individual differences, measurement error, and/or other variables not included in the model. We will talk more about model estimation later on. Recall, however, that we use samples drawn from the population to estimate the model parameters (i.e., μ and α_j).

What distinguishes one-way ANOVA from t tests with independent samples? In other words, if the focus is on comparing means between groups, why not just use the t test for independent samples for each pair of groups? Abu-Bader (2006) identifies two reasons. First, in studies with more than two groups, it would be necessary to conduct multiple t tests in order to examine mean differences among all the possible pairs of groups. For instance, perhaps you wish to conduct an experiment examining an intervention to increase healthy eating among children in low-income families. Previous research suggests that your pretest measure may have an effect on outcomes, independent of the intervention. Thus, you decide to test this intervention using a Solomon Four-Group design so you can examine pretest effects. Your design looks like this:

Experiment 1 Treatment:	R	O_1	X	O_2
Experiment 1 Control:	R	O_3		O_4
Experiment 2 Treatment:	R		X	O_5
Experiment 2 Control:	R			O_6

If you used independent-samples *t* test to examine mean differences between groups, you would have to conduct *six separate tests* to determine differences in the means for all possible pairs (i.e., [a] Experiment 1 treatment vs. Experiment 1 control, [b] Experiment 1 treatment vs. Experiment 2 treatment, [c] Experiment 1 treatment vs. Experiment 2 control, [d] Experiment control 1 vs. Experiment 2 treatment, [e] Experiment 1 control vs. Experiment 2 control, and finally [f] Experiment 2 treatment vs. Experiment 2 control). As you can see, this is labor intensive.

Building on this, the second reason for not using the independent-samples *t* test when the mean of more than one pair of groups is compared is because of the increased risk of making a **Type I error** (i.e., incorrectly rejecting the null hypothesis when it is true). The alpha level for one *t* test is typically set at .05. For a set of *t* tests, the overall alpha level increases making it larger than the alpha level for only one test. This is because the risk of making a Type I error grows or accumulates across the multiple comparisons. Lomax (1992) provides a useful analogy. He states, "For each test we are taking a risk; the more tests we do, the more risks we take. This can be explained by considering the risk you take each day you drive your car to school or work. The risk of an accident is small for any one given day. However, over a period of a year the risk of experiencing an accident is much larger" (p. 89).

What would the Type I error be if we used independent-samples *t* tests to evaluate the differences in means among groups in our Solomon Four-Group example? We would need to conduct six tests. We set an alpha level of .05 for each test, indicating that the probability of falsely rejecting the null hypothesis (Type I error) is 5%, and the probability of not making a Type I error is 95% for each test. Because each test is independent, we multiply their probabilities to determine the overall probability of not making a Type I error, which is $.95^6 = .95 \times .95 \times .95 \times .95 \times .95 \times .95 = .74$. Thus, the probability of making at least one Type I error is .26 or 26% (1 – .74), if we used independent-samples t -tests. So, in this example, the probability of making a Type I error increased from 5% to 26%! The error rate across multiple tests in the same analysis is called the **family-wise (i.e., experiment-wise) Type I error rate.**

In order to correct the inflated error rate in ANOVA, we use what is referred to as an overall **omnibus test** to examine difference in means across multiple groups. This involves conducting only one test that compares mean differences between groups simultaneously. The F statistic can be used for this purpose. In essence it is an extension of the independent t test for two or more sample means that controls for the **family-wise error rate.** We cover the F statistic in more detail in the next section.

Relevant Basic Statistics in ANOVA

Essentially, ANOVA is built on two basic statistics—the mean and the variance. The variance is particularly important because it is used to determine the sums of squares and the omnibus F test. First, however, we discuss the mean. The purpose of using ANOVA in data analysis is to determine differences in means between groups. Thus, the mean is also a very important statistic. Recall that the mean is defined as the arithmetic average of a set of scores in a distribution. It is one of the three measures of central tendency (i.e., mean, median, and mode) used to describe data. The mean is calculated by summing the values of all the scores in a distribution, and then dividing the total by the number of cases in the study. For instance, if the number of cases is 175, and the values of a set of scores sum to 2,051, then the mean is 11.72 (2051 / 175). Because the focus is on group differences across means, the DV in ANOVA models must always be measured at the interval or ratio-level so that a legitimate mean can be determined.

The other important statistic is the **variance.** In ANOVA, estimates of the variance are used to compare means between groups. The variance is one of the three measures used to describe the dispersion, or variability, of the data for a particular variable. Other dispersion measures are the range and the standard deviation. The variance provides an estimate of how spread out the distribution of scores is from the variable's mean (Abu-Bader, 2006). The value of the variance indicates the extent of that spread. The more spread out the scores are from the mean, the larger the value of the variance. The closer the scores cluster around the mean, the smaller the value of the variance. The variance provides information about how well the estimated model actually fits the data. This is because it measures the average error (i.e., spread) between the mean and the scores (Field, 2005).

In ANOVA, the focus is on identifying where variance occurs—between and/or within groups, more formally referred to as **between-groups variance** and **within-groups variance**. Remember that in ANOVA we wish to determine mean differences between the groups in our study. Between-groups variance indicates how the means vary from group to group (e.g., treatment vs. control group) relative to the combined mean for all groups. Within-group variance provides a measure of the variation in mean scores for each participant in the same group (e.g., the mean scores of participants in the treatment group), relative to the group mean. The total variance is the sum of the between-group variation and the within-group variation. In experimental research, we hope to find more dispersion between the treatment and control groups than within each group. This would reflect differences between the groups and, depending on our research design, might indicate the effectiveness of the intervention.

In the previous chapter we covered partitioning the **sums of squares** in *Y*. Recall that the total sum of squares is partitioned into two components—sum of squares predicted by the regression equation (SS_M) and the left over sum of squares (SS_R), represented as $SS_T = SS_M + SS_R$. The total sum of squares represents the amount of total variation in *Y*, part of which includes the variation explained by the model (S_M) and the other part is considered error or residual (S_R).

The total sum of squares is also partitioned into two components in one-way ANOVA. In this case, the total sum of squares (i.e., total variation) is represented by the sum of the variation between the groups, denoted SS_{betw}, and the sum of the variation within the groups, denoted SS_{with}. SS_{betw} is represented as SS_M in linear regression, and SS_{with} is represented as SS_R or the error. Remember, we hope to find variation in the mean differences between groups and little variation (i.e., error) within each group. As you can see, variance is a critical concept in ANOVA.

The variance has an important role in determining statistical significance of ANOVA models. Earlier we mentioned that the omnibus *F* test (rather than multiple *t* tests) is used to determine statistical significance in ANOVA. The *F* test allows us to test for mean differences between groups simultaneously, decreasing the likelihood of incorrectly rejecting the null hypothesis when it is true (i.e., Type I error). The *F* test produces the F statistic, which is the ratio of the two estimated components of the total variance—the between-groups estimated variance and the within-groups estimated variance. The formula is: F = between-groups estimate of variance/within-groups estimate of variance. Because the F statistic has an F sampling distribution, the F-Table is used to determine statistical significance. If the F test is statistically significant, then the null hypothesis that all population means are equal is rejected. This also means that the alternative hypothesis (i.e., at least one population mean is different from the other population means) is retained. Note that this *does not* indicate which particular population mean differs from the means of the other groups. To determine which group mean(s) differs from the other group means, it is necessary to conduct planned or post hoc-comparisons of means, also called contrasts. We cover contrasts later in this chapter.

Let's get back to the F statistic, in particular how its numerator (i.e., between-groups variation) and denominator (i.e., within-groups variation) are estimated. The estimate of the numerator represents the *mean variance of the between-groups sum of squares*. The estimate of the denominator represents the *mean variance of the within-groups sum of squares*. The between-groups mean square, denoted MS_{betw}, is calculated by dividing the sum of the variation between the groups (i.e., SS_{betw}) by its degrees of freedom. The degrees of freedom for the between-groups sum of squares is calculated by subtracting 1 from the total number of groups in the model. If k = the total number of groups in the model, then the formula for the degrees of freedom between groups is $k - 1$. Similarly, the within-groups mean square, denoted MS_{with}, is calculated by dividing the sum of the variation within the groups (i.e., SS_{with}) by its degrees of freedom. The degrees of freedom for the within-groups sum of squares is calculated by subtracting the total number of groups in the model (k) from the total sample size (n), or $n - k$. Thus, the formula for the F statistic is MS_{betw}/MS_{with}, where MS_{betw} is the estimate of the mean between-groups variance and MS_{with} is the estimate of the mean within-groups variance.

The ANOVA Summary Table

How the sources of variation are partitioned in one-way ANOVA is often reported in a table referred to as the **ANOVA Summary (i.e.,**

Table 6.1 One Way ANOVA Source of Variation Table

Source of Variation	*Sum of Squares*	*df*	*Mean squares*	*F Test*
Between groups	SS_{betw}	k–1	$MS_{betw} = SS_{betw}/(k-1)$	$F = MS_{betw}/MS_{with}$
Within groups	SS_{with}	n–k	$MS_{with} = SS_{with}/(n-k)$	
Total	SS_{total}	n–1		

Source) Table. In fact, SPSS provides this table as a part of its output. Table 6.1 provides the general form of the ANOVA summary table. The first column lists the two sources of variation—between-groups and within-groups—and the total variation. The second column provides the sum of squares for each source. The third column indicates their degrees of freedom. The mean squares reported in the fourth column are the estimated average variances. The mean between-groups variance is determined by dividing the between-groups sum of squares by its degrees of freedom (i.e., $k - 1$). The mean within groups variance is calculated by dividing the within-groups sum of squares by its degrees of freedom (i.e., $n - k$). Finally, the F statistic is simply the ratio of the between-groups mean sum of squares and the within-groups mean sum of squares.

Estimating Model Parameters and Measures of Association

Recall that the population equation for ANOVA is $Y_{ij} = \mu + \alpha_j + \varepsilon_{ij}$. Thus, the model parameters are μ, α_j, and ε_{ij}. As in multiple regression methods, we actually estimate a set of sample statistics to represent the population parameters. In ANOVA, the sample statistics are $\bar{Y}..$, α_j, and e_{ij}, where

- $\bar{Y}..$ – Sample grand mean (i.e., the mean across all subjects in the sample)
- α_j = group effect for the jth group; it is computed as $\alpha_j = \bar{Y}_j - \bar{Y}..$; $\bar{Y}_i$ = the sample mean for all i individuals in group j
- e_{ij} = error (i.e., residual) or the difference between the observed scores and what is predicted by the model; it is computed as $e_{ij} = Y_{ij} - \bar{Y}_j$. Note that a large error term indicates that the model

  does not predict the observed scores well, so we hope for a small error term.

Similar to multiple regression analysis, ANOVA model parameters are estimated using the least square estimator. In ANOVA, we also wish to determine a set of estimates that minimizes the sum of the squared residuals. The least squares estimator provides estimates such that the sum of the squared residuals is smallest.

In ANOVA, three measures of effect size are produced to provide information about the association between *X* and *Y*—eta squared (η^2), adjusted η^2, and omega squared (ω^2). Eta squared (η^2) indicates the portion of variance in *Y* explained by *X* (i.e., differences between groups). It is a correlation ratio that is calculated by dividing SS_{betw} (between-groups sum of squares) by SS_{total} (total sum of squares) (i.e., SS_{betw}/SS_{total}). Eta squared is similar to the R^2 statistic in multiple linear regression analysis. Like R^2, η^2 is positively biased in that it tends to overestimate the association between *X* and *Y*, especially with small samples (i.e., < 30). Thus, SPSS provides an adjusted η^2 that is not affected by sample size. The other effect size measure in ANOVA is omega squared (ω^2). This measure tends to be more accurate than η^2 in that it is not positively biased, but it requires the same number of participants in each group. In any case, CIs can be calculated to assess the precision of these estimates relative to the population values.

Assumptions and Other Model Requirements

As with multiple linear regression analysis, one-way ANOVA produces estimates of parameters that are intended to represent a population. ANOVA is a parametric test and thus shares the four assumptions described in Chapter 4—DV measured at the interval or ratio-level data, normally distributed data, homogeneity of variance, and independence of observations—and the three assumptions about the distribution of the residuals or errors (i.e., ε_{ij})—that they are normally distributed, independent within and across groups, and with equal variance. We can summarize this into one statement—errors are assumed to be independent and normally distributed with equal variances. In the following section, we review the three assumptions about the residuals and what to do if the assumptions are violated.

First, errors are assumed to be independent within and across groups. In other words, the scores of one participant should not influence the scores of another participant. The pattern of errors should be random rather than systematic. Systematic error might be indicated when the errors in one group are large compared to the errors of the other groups, suggesting a relationship between X and ε (Lomax, 1992). A violation of this assumption may result in incorrect standard errors of the sample means, which can increase the likelihood of making a Type I or Type II error. Potential violations can be assessed by examining the **residual plots** of the groups. No discernable pattern of data points should be observed. The Durbin–Watson test can also be used to evaluate violations of the independence of error assumption. Time series designs, designs that involve observations of groups (e.g., multiple classrooms of students), and designs with repeated observations are particularly vulnerable to violations of the independence assumption. On the other hand, this assumption is usually met when participants are randomly assigned to groups or with designs that include only one observation.

Second, errors are assumed to be normally distributed. Graphs such as histograms, stem and leaf plots, or box plots can be used to determine whether the distribution of the errors is shaped in a way that resembles a normal curve. As mentioned in Chapter 5, results remain fairly robust in the context of moderate violations of the normality assumption. Again, in severe cases, you can consider using nonparametric tests that do not require a normal distribution of errors, or transforming the shape of the distribution.

The final assumption is that the distributions of the residual errors have equal variances across groups. This is referred to as **homogeneity of variance** or **homoscedasticity**. A violation of the homogeneity assumption may lead to an incorrect SS_{with} (i.e., within-groups sum of squares) which increases the likelihood of making Type I or Type II errors. As with violations of the normality assumption, violations of the homogeneity of variance assumption are less likely to occur with large samples and when sample sizes across groups are equal. One way to evaluate whether this assumption has been violated is to compare the variances of each group—they should be fairly close in size. By default, SPSS reports the Levene test as an indicator of whether the homogeneity of variance assumption was met. Other tests are the Cochran's C and the Hartley F_{MAX}.

Multiple Contrasts

So far we've covered statistical techniques that focus on overall results (i.e., omnibus F test, η^2, adjusted η^2, and ω^2). Researchers are also interested in assessing which of the means between pairs of groups is significantly different from one another. This is determined by conducting **planned** or **posthoc comparisons** or **contrasts** (denoted in the population as ψ). In this context, the terms "contrasts" and "comparisons" are often used interchangeably. Lomax (1992) refers to a contrast as "a weighted combination of the means" (p. 127), that entails comparing the means between two groups (e.g., group 1 mean vs. group 2 mean), referred to as a simple contrast, or sets of groups (e.g., group 1 and group 2 means vs. group 3 mean), referred to as a complex contrast. Regardless of the total number of groups, contrasts are always pairwise, or made between two components or entities. The purpose of conducting contrasts is to test specific hypotheses about differences between group means.

Procedures for conducting contrasts in ANOVA are designed to counteract the likelihood of making a Type I error (i.e., incorrectly rejecting a true null hypothesis) in comparing multiple means across groups. These methods minimize the impact of family-wise error rates. Thus, we can conduct tests to compare means between groups in our study (as if conducting several *t* tests) in a way that strengthens the likelihood that the family-wise error rate does not rise above .05.

Two types of contrasts are available—planned and post-hoc. Planned contrasts are hypothesized *a priori*, based on theory or previous research. Post-hoc contrasts are done after examining initial model results, often in response to a significant omnibus *F* test. Note that most contrast procedures require independent and normally distributed residuals with equal variances. Some procedures require additional assumptions such as equal sample sizes across groups. In the following section, we provide an overview of common planned and post-hoc contrast tests that are available in SPSS. First, we explain in more detail the nature of these tests.

Planned Contrasts

Planned contrasts are conducted to test specific hypotheses about group-mean differences developed *a priori*. These hypotheses are derived before data are analyzed, and are based on theory or previous research. Thus, compared to post-hoc contrasts, planned contrasts tend to be more specific and focused. Usually, fewer planned contrasts are conducted. The statistical significance of the overall *F* test is of little relevance in conducting planned contrasts.

Recall that, in ANOVA, the total variance is split into two parts—between-groups variance and within-groups variance. In the context of an experimental design, between-groups variance represents variation due to the experiment (SS_{betw} or SS_M), whereas within-group variance represents variation due to error, or unexplained variance (SS_{with} or SS_R). The focus of planned contrasts is to examine variation due to the experimental manipulation (SS_{betw}). The variance attributed to the experiment, SS_{betw}, is partitioned into components. We then examine contrasts between these components. For instance, say our experiment involved three groups—a group that got the intervention (i.e., treatment group), a group that got services as usual (i.e., alternative treatment group), and a group that got no services (i.e., control group). Two sets of contrasts would be tested to determine how the variance explained by the experiment (i.e., between-groups variance) was partialled out. In the first contrast, we might compare mean differences between the groups that got services (i.e., intervention group and services as usual group) and the group that got no service. The second contrast would then involve examining mean differences between the services groups—the intervention group and the services as usual group. As adapted from Field (2005, p. 326), Figure 6.1 illustrates this.

Three rules guide the process of conducting planned contrasts. First, each contrast must include only two components. It may be that one or more of the components include multiple groups, as in the previous example. The second rule is that component parts (i.e., groups in the study) are mutually exclusive and independent of one another. In other words, a group is only included *once* in a contrast. Third, the total number of contrasts tested is always one less than the total number of groups in the study. The previous example includes three groups (i.e., treatment group, services as usual group, and control group) and two contrasts.

Contrasts must be defined prior to running the tests. This is done by assigning values to the coefficients of the equations that represent the contrasts. These values are referred to as **weights**. Procedures for assigning weights depend on whether a set of contrasts is **orthogonal** or **nonorthogonal**. The following section describes what is meant by orthogonal and nonorthogonal contrasts, and provides rules for assigning weights to contrasts.

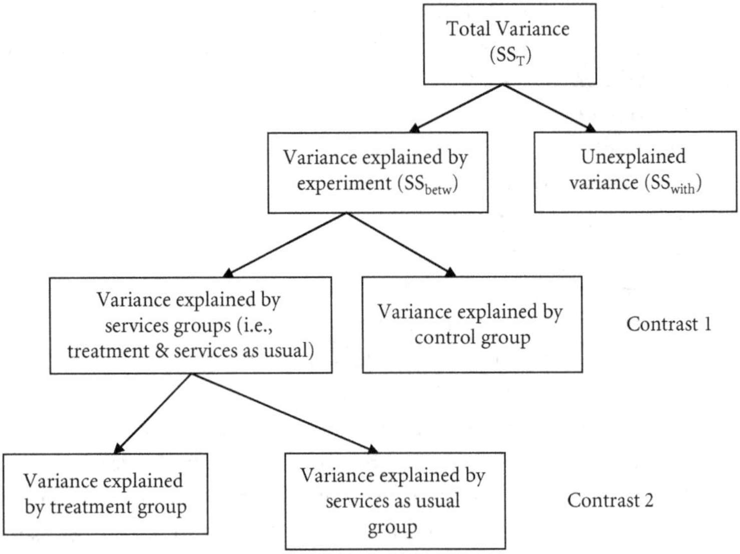

Figure 6.1 Partitioning the total variance of a three-group experiment into explained and unexplained variance, and then into two contrasts. Adapted from Discovering Statistics using SPSS, by A. Field, 2005, p. 326, Sage Publications. Copyright 2005 by Sage Publications.

Orthogonal Contrasts

A set of contrasts is orthogonal if the component variances are independent of one another. Orthogonality is based on the sum of the values of contrast weights, multiplied together. If the products of the weights in a set of contrasts sum to zero, then these contrasts are considered orthogonal, and thus independent. Orthogonal contrasts require that sample sizes across groups are equal. We will discuss how to conduct contrasts in cases where group sample sizes are not equal later in the chapter.

Table 6.2 provides an example of a set of orthogonal contrasts with weights. The example is based on three groups with three means (i.e., μ_1, μ_2, and μ_3). We assign three weights or coefficients (C_1, C_2, and C_3) to each mean. The first column shows two sets of contrasts (i.e., ψ_1 and ψ_2), one less than three, the total number of groups in the study. The values of the weights for each contrast (shown in the third and fourth rows) differ from one another. For example, the values of the weights for the

Table 6.2 Orthogonal Contrasts

	Weight (i.e., Coefficient) Value		
Contrast	C_1	C_2	C_3
$\psi_1: \mu_1 - \mu_2 = 0$	+1	−1	0
$\psi_2: 1/2\,\mu_1 + 1/2\,\mu_2 - \mu_3 = 0$	+1/2	+1/2	−1
Total	1 × 1/2 = 1/2	−1 × 1/2 = −1/2	−1 × 0 = 0

first contrast (ψ_1) are +1, −1, and 0. The value of the third weight in this contrast is 0 because the mean of the third group is not being evaluated at this point. To determine whether this set of contrasts is orthogonal, we multiply the values of the weights down each of the second, third, and fourth columns and then add them together. If they sum to zero, then the set of contrasts is orthogonal. As you can see, the column totals [i.e., ½ + (−½) + 0] sum to zero, which means this set of contrasts is orthogonal.

To summarize, Field (2005) identifies the following rules to assign values to weights.

- Rule 1: Each contrast includes only two components of variation.
- Rule 2: In the first comparison, combine all experimental groups (e.g., treatment and services as usual groups) to compare them with control group(s).
- Rule 3: If a group is singled out in one comparison, that group is then excluded from subsequent contrasts.
- Rule 4: Assign positive values of weights to one component of the contrast and negative values of weights to the other component.
- Rule 5: If a group is not included in the comparison, assign it a weight of zero.
- Rule 6: The sum of weights for a comparison equals zero. (p. 330)

SPSS offers the Planned Orthogonal Contrast (POC) procedure to conduct a test of orthogonal contrasts. This test is used when contrasts have been defined *a priori* (i.e., planned rather than post-hoc) and the set of contrasts is orthogonal (i.e., mutually exclusive). Note that sample sizes for each group must be equal when using the POC procedure. If not, then you can use the Tukey–Kramer test for unequal sample sizes, which is discussed later in the chapter.

Nonorthogonal Contrasts

It is also possible to conduct tests of a set of nonorthogonal contrasts. Nonorthogonal contrasts include components that are not mutually exclusive. The sources of variation overlap or are not independent of one another. Nonorthogonal contrasts differ from orthogonal contrasts in that they do not represent nonredundant and independent sources of variation. In this case, it is necessary to use a more conservative probability level to determine the statistical significance of a particular contrast. The Bonferroni Correction (discussed later) can be used to correct for the overlap. Also, interpreting the results of nonorthogonal contrasts is not as straightforward, relative to orthogonal contrasts.

Table 6.3 provides a set of nonorthogonal contrasts, based on a simple modification of Table 6.2. As you can see, the example still includes three groups, each with its own mean (i.e., μ_1, μ_2, and μ_3) and three weights or coefficients (i.e., C_1, C_2, and C_3) that have been assigned values. Two contrasts are examined (i.e., ψ_3 and ψ_4). The first contrast is the same as what was shown in Table 6.2. However, in the second contrast the means of only two groups, 1 and 3, are compared, instead of comparing means of all three groups as we did in the previous example. Thus, the values of the weights as well as their products are different. As you can also see, the products sum to 1 (1 + 0 + 0) rather than 0, which indicates that this set of contrasts is not orthogonal.

Post-Hoc Contrasts

Post-hoc contrasts are similar to planned contrasts in that the intent is to compare means of different combinations of groups in the study. There are some critical differences however. Post-hoc contrasts are conducted when no specific hypotheses have been posited *a priori*. The statistical significance (or not) of the overall *F* test informs the decision to conduct

Table 6.3 Nonorthogonal Contrasts

	Weight (i.e., Coefficient) Value		
Contrast	C_1	C_2	C_3
$\psi_3: \mu_1 - \mu_2 = 0$	+1	−1	0
$\psi_4: \mu_1 - \mu_3 = 0$	+1	0	−1
Total	$1 \times 1 = 1$	$-1 \times 0 = 0$	$-1 \times 0 = 0$

post-hoc contrasts. They should only be conducted when the overall *F* test is statistically significant (Howell, 2010). In post-hoc contrasts, all possible pairs of means are tested. Because of this, the possibility of an inflated Type I error rate increases. The tests for conducting post-hoc contrasts correct for this by adjusting the Type I error critical value (α). These adjustments are intended to control the family-wise error rate. Different types of adjustments are made depending on the specific test. Similar to planned comparisons, most post-hoc tests require independent and normally distributed errors with equal variances.

A number of tests are available to conduct post-hoc contrasts, each with different ways of controlling the family-wise error rate. Four post-hoc tests that are frequently conducted when equality of variances is assumed include Fisher's LSD (least significant difference), Bonferroni (also known as the Bonferroni correction), Scheffe, and Tukey's Honestly Significant Difference (HSD) test. These tests require equal sample sizes across groups. When sample sizes differ between groups, one can use the Tukey Kramer test. When the homogeneity of variance assumption is not met, the Tamabane's T2 can be used. Finally, the Kruskal-Wallis test is a non-parametric test that can be used when the normality assumption is violated. For a comprehensive presentation of these tests (and others) see Lomax (2007).

Examples of Social Work Research using One-Way ANOVA

ANOVA procedures are used to examine questions about differences in means between groups. Thus, ANOVA procedures are common in experimental research where the focus is on determining the effects of an intervention by comparing outcomes of treatment and control groups. For instance, Edmond, Rubin, and Wambach (1999) used ANOVA procedures to assess the impact of eye movement desensitization and reprocessing (EMDR) in reducing trauma among adult females who experienced childhood sexual abuse. Fifty-nine participants were randomly assigned to three groups—EMDR treatment ($n = 20$), routine treatment ($n = 20$), and control ($n = 19$). Mean differences between groups on several outcome measures were compared. Outcomes (i.e., the DVs) were measured on an interval scale. The IV (i.e., assignment into the intervention group, routine treatment group, or control group) was measured on a nominal scale with three categories.

ANOVA procedures can also be used to analyze data from studies that use quasiexperimental designs. Painter (2009) used a pre- and posttest quasiexperimental design to compare multisystemic therapy (MST) to the "services-as-usual" family skills training among a group of emotionally disturbed children. Data were collected retrospectively from a community mental health services computerized data base of clients who had completed MST ($n = 87$) or services as usual ($n = 87$). One-way ANOVA was used to compare mean differences on improved functioning, using responses to the Ohio Problem Severity scale (Ogles, Lunnen, Gillespie, & Trout, 1996) and the Ohio Youth Functioning scale (Ogles et al.). Results showed that youth who completed MST scored higher on outcomes, relative to the services-as-usual group. In this example, the DVs (the Ohio Problem Severity scale and the Ohio Youth Functioning scale) were measured on an interval scale. The IV had two categories (MST group and services-as-usual group) and was measured on a nominal scale.

TWO-WAY ANALYSIS OF VARIANCE (ANOVA): A DETAILED DESCRIPTION

In one-way ANOVA models, cases are separated into groups based on *one factor* such as race/ethnicity, or in the case of experimental designs, one of the study's conditions (e.g., treatment, services-as-usual, or control). This allows us to compare means of multiple groups across one factor. It is often the case that we are interested in comparing means across more than one factor (e.g., study conditions AND race/ethnicity). In other words, we wish to examine the impact of multiple IVs on the DV. **Two-way ANOVA** allows us to do just that for designs with two factors (i.e., IVs). At the beginning of the chapter, we described a study conducted by Lindsey and others (2010) to evaluate the Family Connections program. The authors used ANOVA procedures to assess differences in outcomes between groups randomly assigned to one of four study conditions. The factor or IV in this case was "study condition." Suppose they were also interested in differences across the four study conditions based on family structure (e.g., two-parent families or single-parent families), thus introducing a second IV, family structure, with cases separated into two groups—two-parent and single-parent families. The researchers could use two-way ANOVA to determine mean differences across both study conditions and family structure.

Much of what was presented on one-way ANOVA is applicable in two-way ANOVA models. There are some differences, however, which are described in the following section and organized as follows. First, we define two important concepts—main effects and interaction effects—upon which two-way ANOVA models are based. Next, we present the components of the two-way ANOVA population model, explain estimation procedures, and describe how two-way ANOVA sums of squares differ from one-way ANOVA. We then describe the way in which two-way ANOVA assumptions differ from the assumptions of one-way ANOVA. We build on the discussion of multiple comparisons in one-way ANOVA to provide a brief overview of how this is done in two-way ANOVA. Finally, we provide examples of social work research that have used two way ANOVA as the data analysis method.

Important Concepts

Two-way ANOVA procedures involve comparing means between multiple groups on two factors for both main effects and interaction effects. A **main effect** is the direct or unique impact of a factor (i.e., IV) on the DV. We test for main effects in one-way ANOVA by determining mean differences between groups. In the case of two-way ANOVA, we determine the main effects of two factors, A and B, separately on the DV. In other words, we determine the unique effect of factor A on the DV, controlling statistically for factor B, and the unique effect of B on the DV, controlling for A. Building on our previous example of the Family Connections evaluation (Lindsey et al., 2010), factor A would include the four study conditions (i.e., 3-month regular, 9-month regular, 3-month enhanced, and 9-month enhanced). Factor B would be family structure (i.e., two-parent and single-parent families). We evaluate study conditions and family structure for their main effect, independent of the other factor. The evaluation of each factor on the group means (the DV) produces an F statistic as an indicator of whether the group means are equal.

An **interaction effect** represents the combined effect of the two factors on the group means. The purpose of testing for an interaction effect is to determine whether the two factors operate independent of one another (no interaction exists) or whether the two factors work together to produce an additional impact, beyond the impact of their main effects. In our example, we would examine whether study conditions and family structure combine to have an impact on the outcomes, in addition to their main effects. For instance, it may be that a longer intervention is more successful for single-parent families that may not have as many resources as two-parent families. We could determine this by testing the interaction effect of study condition and family structure on group mean outcomes. As with main effects, an F statistic is produced as an indicator of statistical significance of the interaction. There are many ways in which factors can combine with one another to produce interaction effects. Lomax (1992, pp. 168–170) provides a detailed description for those who would like additional information.

Two-Way ANOVA Population Model

Recall that the equation of the one-way ANOVA model is: $Y_{ij} = \mu + \alpha_j + \varepsilon_{ij}$, with one factor A. The two-way ANOVA model has two factors, A and B. The formula is: $Y_{ijk} = \mu + \alpha_j + \beta_k + (\alpha\beta)_{jk} + \varepsilon_{ijk}$, where,

- Y_{ijk} = score on the DV for individual i in level j of factor A and level k of factor B (or the jk cell),
- μ = population grand mean,
- α_j = main effect for level j of factor A,
- β_k = main effect for level k of factor B,
- $(\alpha\beta)_{jk}$ = interaction effect for level j of factor A combined with level k of factor B, and
- ε_{ijk} = error for individual i in cell jk

In two-way ANOVA, the intent is to explain *more* of what we referred to as "error" in the one-way ANOVA model by determining the impact of a second factor B and the interaction of A and B on the DV, in addition to the impact of factor A. Hopefully we can explain more variance as indicated by η^2 and thus reduce the amount of error in the model.

It is useful to tease out the notation for Y_{ijk} to understand what is meant by "cell jk." The j subscript indicates a particular level of factor A for the ith observation, and the k subscript indicates a particular level of factor B for the ith observation. The ith observation represents a particular case in the dataset. Each case or observation has its own Y_{ijk} notation, depending on what observation it is, from which level of factor A and from which level of factor B. Y_{412}, for example, is the fourth observation

Table 6.4 Representation of Two-Way (i.e., Factor) ANOVA

Treatment Conditions	*Family Structure*		*Treatment condition marginal means*
	Two-parent	*Single parent*	
3-month enhanced	Cell Mean Y_{11}	Cell Mean Y_{12}	Marginal Mean Y_{1}
9 month enhanced	Cell Mean Y_{21}	Cell Mean Y_{22}	Marginal Mean Y_{2}
3 month regular	Cell Mean Y_{31}	Cell Mean Y_{32}	Marginal Mean Y_{3}
9 month regular	Cell Mean Y_{41}	Cell Mean Y_{42}	Marginal Mean Y_{4}
Family structure marginal means	Marginal Mean $Y_{.1}$	Marginal Mean $Y_{.2}$	Overall Mean Y

in the first level of factor A and the second level of factor B (Lomax, 1992, p. 163).

This idea is displayed in Table 6.4, using the factors from the modified version of the Lindsey et al. (2010) study. The levels of one factor are included as rows (in this case four treatment conditions in rows 3 through 6) and the levels of the other factor are included as columns (in this case two types of family structure in columns 2 and 3). As shown, the rows and columns intersect at several points to form cells. Each cell is made up of the mean scores from individuals who belong to the corresponding levels of each factor. The column means for each factor level, referred to as **marginal means,** are provided at the bottom of the respective column. At the end of each row are the marginal means for each level of the factor displayed in that row. The value in the lower right corner is the overall mean. Each cell provides the mean of all the scores within the cell. The subscript numbers and dots after each cell and marginal mean represent their location in the table. The first number or dot is always associated with the row location and the second number or dot is always associated with the column location. For example, the mean score for all two-parent families in the 3-month enhanced treatment condition would be reflected in the first cell of the table, mean of Y_{11}.

Estimating Two-Way ANOVA Models

The least squares method is used to estimate the population parameters from the sample data in two-way ANOVA models. Table 6.5 provides the formulas and hypotheses for testing each estimate.

Table 6.5 Two-Way ANOVA Model Estimation Formulas and Hypotheses

Population Parameter	*Sample Estimate*	*Formula*	*Hypotheses*
α_j	a_j	$\bar{Y}_{.j.} - \bar{Y}\ldots$	H_{01}: $\mu_{.1.} = \mu_{.2.} = \mu_{.3.}\ldots = \mu_{.j.}$ H_{A1}: not all the $\mu_{.j.}$ are equal
β_k	b_k	$\bar{Y}_{..k} - \bar{Y}\ldots$	H_{02}: $H_0 = \mu_{..1} = \mu_{..2} = \mu_{..3}\ldots = \mu_{..k}$ H_{A2}: not all the $\mu_{..k}$ are equal
$(\alpha\beta)_{jk}$	$(ab)_{jk}$	$\bar{Y}_{.jk} - (\bar{Y}_{.j.} + \bar{Y}_{..k} - \bar{Y}_{jk}\ldots)$	H_{03}: $(\mu_{.jk} - \mu_{.j.} - \mu_{..k} + \mu) = 0$ for all j and k H_{A3}: not all the $(\mu_{.jk} - \mu_{.j.} - \mu_{..k} + \mu) = 0$
ε_{ijk}	e_{ijk}	$\bar{Y}_{ijk} - \bar{Y}_{.jk}$	

Note. $\bar{Y}\ldots$ is the overall sample mean; $\bar{Y}_{.j.}$ is the sample mean for level j of factor A; $\bar{Y}_{..k}$ is the sample mean for level k of factor B; $\bar{Y}_{.jk}$ is the sample mean for cell jk; and the dot subscripts indicate what effect each mean represents (i.e., i, j, and/or k). Adapted from *Statistical Concepts*, by R. G. Lomax, 1992, pp. 164–167, Longman Publishing Group. Copyright 1992 by Longman Publishing Group.

Sums of Squares in Two-Way ANOVA

The way in which the sums of squares are partitioned in two-way ANOVA builds on what has already been presented for one-way ANOVA. In one-way ANOVA the total amount of variation in the model (SS_{total}) is partitioned into two components—**between-groups sum of squares** (SS_{betw}) and **within-groups sum of squares** (SS_{with}). SS_{with} represents the error in the model. In two-way ANOVA, the total sum of squares (SS_{total}) also represents the amount of total variation and the within-groups sum of squares, SS_{with}, represents the error. What is different in two-way ANOVA is that the between-groups sums of squares component is expanded to account for the additional factor B and the interaction term. The notations for these sums of squares are SS_A, SS_B, and SS_{AB}. The equation is: $SS_{total} = SS_A + SS_B + SS_{AB} + SS_{with}$. SS_A is the variation explained by factor A; SS_B is the variation explained by factor B; and SS_{AB} is the variation explained by the interaction of factors A and B. In two-way ANOVA we attempt to explain more of the total variation with the inclusion of a second factor and the interaction between the two factors, relative to one-way ANOVA. We would expect the value of the within sum of squares, SS_{with}, to be smaller in two-way ANOVA models relative to one-way ANOVA models because the second factor and the interaction term should hopefully explain more of the total variation.

Assumptions

Similar to one-way ANOVA, in two-way ANOVA we assume that errors are independent and normally distributed with a common variance. In this case, however, the focus is on the residuals within each cell. These residuals reflect the difference between the values of each observation and the cell mean. To determine error for each ith case, we subtract its score from the cell mean. The difference represents error. The consequences of violating assumptions in two-way ANOVA are similar to the consequences for violating the assumptions in one-way ANOVA. The same techniques can be used to detect assumption violations. However, the residual plots of cells rather than of the groups are examined.

Multiple Comparisons in Two-Way ANOVA

In two-way ANOVA, contrasts of both main effects and interaction effects are conducted. Much of what was presented in the previous section on conducting multiple contrasts for one-way ANOVA can be applied to conducting contrasts in two-way ANOVA. What is different here is that in two-way ANOVA models, we are interested in examining contrasts of both main and interaction effects. Thus, two-way ANOVA models include more possible means to compare (i.e., cell and marginal means as shown in Table 6.4). It is important to clearly indicate which means are being compared. As sample sizes within each cell and for each marginal mean are likely to vary, it is also important to clearly indicate the sample size for the particular means being compared. The process for conducting contrasts in two-way ANOVA models is beyond the scope of this book. If you are interested in this topic, see Field (2005) or Lomax (1992).

Examples of Social Work Research using Two-Way ANOVA

Two-way ANOVA procedures are used when researchers wish to compare means across two factors. In other words, we examine the impact of two IVs, measured categorically, on one DV, measured at the interval or ratio-level. In their randomized control trial, Silver and Bauman (2006) compared differences in HIV/AID knowledge and sexual attitudes, intentions, and peer norms between two groups of inner city youth. Two-way

ANOVA procedures were used to examine mean differences in three outcomes (i.e., HIV/AID knowledge, abstinence self-efficacy, and outcome expectancies) based on sexual experience and gender. Thus, in this study the two factors or categorical IVs were sexual experience and gender, each with two levels—sexual experience or sexual inexperience, and male or female, respectively. The outcomes (i.e., DVs) were measured at the ratio-level.

Nakonezny, Rodgers, and Nussbaum (2003) employed a quasi-experimental pre- and posttest design with two-way ANOVA methods to examine the impact of later-life parental divorce on relationships between adult children and their parents. This study provides an example of the use of two-way ANOVA procedures in nonintervention research. The sample was divided into divorced families and intact families, and then further divided into another two groups based on predivorce solidarity in the adult–child/older parent relationship. Thus, the two factors are 1) older-parent divorce status with two levels (divorced versus intact) and 2) predivorce solidarity in the adult–child/older parent relationship, also with two levels (high pretest solidarity or low pretest solidarity). Three types of relationship solidarity outcomes were examined—associational solidarity, affectional solidarity, and functional solidarity—all measured at the ratio-level.

ONE-WAY ANALYSIS OF COVARIANCE (ANCOVA): A DETAILED DESCRIPTION

We can build on what we learned about one-way ANOVA to understand ANCOVA procedures. Essentially these procedures are similar, with one major difference. Recall that the purpose of one-way ANOVA analysis is to compare differences in *means* across groups. Thus, the DV is measured at the interval or ratio-level (so we can calculate a legitimate mean) and the factor or IV (the group variable) is categorical. These ideas extend to ANCOVA models as well. However, ANCOVA allows us to examine mean differences between groups, controlling for another variable that may explain some of the reasons for the mean differences between the groups. This is done by adding what is referred to as a **covariate** to the model and testing its impact. Covariates are not DVs or IVs. They are simply variables that are hypothesized to influence mean differences between groups in addition to the IV. By adding the covariate, we expect to explain more variation in the outcome, relative to one-way ANOVA. ANCOVA calculates mean differences between groups, adjusted for the impact of the covariate. Covariates are usually measured at the interval or ratio-level.

Let's use the Lindsey et al. (2010) study to consider an example. Recall that the purpose of this study was to test the impact of Family Connections on child outcomes. The DV was based on responses to the Child Behavior Checklist (CBCL) and the IV was group assignment to one of the four study conditions. Suppose that, in the authors' review of previous research, they discovered that family-based interventions are more successful for younger children. If this is the case, Lindsey and others might include in their model a variable that measures child age, to control or account for differences in intervention outcomes based on children's ages.

ANCOVA Population Model

The equation of the one-way ANOVA model is: $Y_{ij} = \mu + \alpha_j + \varepsilon_{ij}$, with one factor A. In ANCOVA, we add a covariate. The formula is: $Y_{ij} = \mu_Y + \alpha_j + \beta_W (X_{ij} - \mu_x) + \varepsilon_{ij,}$ where,

- Y_{ij} = The outcome score for the ith person in the jth group,
- μ_Y = overall or grand population mean for Y,
- α_j = the group effect for the jth group with the impact of the covariate held constant,
- β_W = the within groups slope from regressing Y on X (i.e., the predicted change in Y given one unit change in X, with group membership held constant),
- X_{ij} = the covariate score of the ith person in the jth group,
- μ_x = overall or grand population mean for X (the covariate), and
- e_{ij} = error (i.e., residual) or the difference for the ith person in the jth group

The Mean and the Variance in ANCOVA

Let's discuss the mean and the variance in the context of ANCOVA. As such we will review three relevant concepts—**covariate** as a source of variation in the model, **adjusted means**, and **partitioning the variance** as represented through sums of squares. A covariate represents a source of variation that has not been included in the experiment, but the evidence (i.e., previous research or theory) suggests that it is likely to affect the DV. Including a meaningful covariate in the model will hopefully increase the amount of total variance that is explained. Lomax (1992) identifies four reasons for including a covariate in an ANCOVA model—(1) to reduce within-group error variance, (2) to account for any preexisting mean group differences based on the covariate, (3) to account for the relationship between the covariate and the DV, and (4) to obtain a more precise and less biased estimate of the effect of the IV on the DV (p. 281).

The group means are adjusted as a consequence of including a covariate in the model. That is, by including the covariate in the analysis, we statistically remove or control for the effects of its variation on the group means of the DV, referred to as adjusted means. In the context of experimental research, the need to adjust group means may occur when one is unable to use random assignment to form the intervention and control groups. Without the use of random assignment, preexisting group differences are likely to be present, and thus confound the relationship between the treatment and the outcome. These preexisting group differences can be represented in the statistical models as confounding variables, or covariates, so their influence can be accounted for.

We can understand how the total variance is partitioned in an ANCOVA model by examining the sums of squares. There are three sources of variation in a one-way ANCOVA model—between-groups adjusted, within-groups adjusted, and the covariate. Note that the between groups and within groups variation have been adjusted to account for the covariate. The equation is: $SS_{total} = SS_{betw(adj)} + SS_{with(adj)} + SS_{cov}$.

ANCOVA Assumptions and Conditions

The assumptions about the residuals in one-way ANOVA models—independence, normal distribution, and common variance—are also assumed for the ANCOVA residuals. ANCOVA includes four additional assumptions related to the covariate, as follows:

- Covariates are measured without error. In other words they are perfectly reliable.

- There is a linear relationship between the covariate and the DV. This is the same assumption we made about the relationship between X and Y in linear regression.
- The covariate relates to the DV in exactly the same way for all groups in the model. This assumption is referred to as **homogeneity of regression.** What it means is that the slopes of the regression lines for each group are equal and the lines are parallel to one another.
- There is no relationship between the covariate and the treatment. In other words, the IV and the covariate are independent of one another. For instance, if group assignment to treatment and control groups was based on age, then age would be an inappropriate covariate because it is likely to be related to the treatment.

SPSS provides several methods for determining whether these assumptions are met. Field (2005) provides a detailed explanation of these methods.

Research Examples Using ANCOVA

Several social work researchers have used ANCOVA as their data-analysis method. Arnold, Smith, Harrison, and Springer (1999) used ANCOVA procedures in their evaluation of the Education Now and Babies Later program. The purpose of this program was to prevent teen pregnancy through abstinence. The evaluation sought to determine whether adolescents who completed the pregnancy-prevention curriculum achieved higher scores on an instrument that measured participant changes in knowledge, beliefs, and attitudes about sexuality. The DV was participant posttest scores, the factor (or IV) was receipt of the intervention or not, and the covariate was participant pre-test scores on the instrument. The use of ANCOVA allowed the researchers to determine mean differences between the treatment and comparison groups, holding constant participant knowledge, beliefs, and attitudes about sexuality prior to the intervention. In other words, the posttest means reflected an adjustment of the impact of the pretest means. Note that with this type of design (i.e., quasiexperimental) the independence of observations assumption does not apply—data were collected from each sample at two time points (i.e., pre- and posttest).

In another example, Auslander and others (2000) used ANCOVA procedures to assess the impact of an intervention designed to reduce the risk of diabetes among low-income African American women. In this case, several outcomes (DVs) related to knowledge, attitudes, and behaviors of healthy eating were assessed. The factor or IV was participation in the intervention or not and the covariate was, once again, the pre-test scores. Mean differences between the treatment and control groups were calculated on all outcome measures, controlling for pretest scores on the measures.

CONDUCTING ONE-WAY ANOVA IN SPSS: AN EXAMPLE

To summarize, one-way ANOVA is used to examine mean differences between two or more groups. It is a bivariate test with one IV and one DV, with the IV measured at the nominal level and the DV measured at the interval or ratio-level. The following provides instructions for conducting one-way ANOVA, and planned and post-hoc comparisons. The instructions also include a check for whether the homogeneity of variance assumption has been met. A more detailed set of instructions is available on the companion website of this book.

The DV is a measure of base year standardized reading scores (BY2XRSTD); the IV is an indicator of the geographic region of each school (G8REGON). The IV has four values—northeast (coded as 1), north central (coded as 2), south (coded as 3), and west (coded as 4). This analysis is based on 300 randomly selected cases with no missing observations from the NELS database. Although not used in this particular analysis, the data set also includes one weight variable (F2PNWLWT).

Instructions for Conducting One-way ANOVA

- Open the SPSS data file entitled, ONE WAY ANOVA.sav.
- Click on "Analyze" in the menu bar at the top of the data view screen.
- A drop down menu will appear. Move your cursor to "Compare Means."

- Another drop down menu will appear to the right of the first menu. Click on "One-way ANOVA" to open the One-way ANOVA dialog box.
- Click to highlight BY2XRSTD in the box on the left side. Click the top arrow to move it to the Dependent List.
- Click to highlight G8REGON and use the bottom arrow to move it to the Factor box.
- Click "Options" in the lower right corner to open the One-Way ANOVA: Options dialogue box. Check "Homogeneity of variance test."
- Click "Continue" to return to the One-Way ANOVA dialogue box.
- Click OK at the bottom of the One Way ANOVA dialogue box to run the one way ANOVA.

Instructions for Conducting Planned Comparisons

- To conduct planned comparisons, click on "Analyze" in the menu bar. Select "Compare Means." Click on "One-way ANOVA" to open the One-way ANOVA dialog box.
- Click to highlight BY2XRSTD and move it to the Dependent List. Click to highlight G8REGON and move it to the Factor box.
- Click "Contrasts" in the upper right corner to open the One-Way ANOVA: Contrasts dialogue box. Move your cursor to the "Coefficients" box. Type "-3" in the box; click "Add"; type "1" in the box; click "Add"; type "1" in the box again and click "Add." Type "1" in the box one more time and click "Add."
- Click "Continue" to return to the One-Way ANOVA dialogue box.
- Click OK at the bottom of the One-Way ANOVA dialogue box to run the one way ANOVA, including planned comparisons.

Instructions for Conducting Post-Hoc Comparisons

- To conduct post hoc comparisons using the Tukey HSD test, click on "Analyze" in the menu bar. Select "Compare Means." Click on "One-way ANOVA" to open the One-way ANOVA dialog box.

- Click to highlight BY2XRSTD and move it to the Dependent List. Click to highlight G8REGON and move it to the Factor box.
- Click "Post hoc" in the upper right corner to open the One-Way ANOVA: Post Hoc Multiple Comparisons box. Check "Tukey."
- Click "Continue" to return to the One-Way ANOVA dialogue box.
- Click OK at the bottom of the One-Way ANOVA dialogue box to run the one way ANOVA, including the post-hoc comparison.
- NOTE: A copy of the output is available on the companion website. The file is entitled, "One Way ANOVA results.spv."

ADDITIONAL READING

Field, A. (2009). *Discovering statistics using SPSS* (3rd ed.). Thousand Oaks, CA: Sage Publications.

Howell, D. C. (2010). *Statistical methods for psychology* (7th ed.). Belmont, CA: Cengage Wadsworth Publishers.

Huitema, B. E. (2011). *The analysis of covariance and alternatives* (2nd ed.). Hoboken, NJ: John Wiley & Sons, Inc.

Lomax, R. G. (2007). *Statistical concepts: A second course* (3rd ed.). New York: Routledge Publishers.

7

Path Analysis

INTRODUCTION

Path analysis is a set of statistical procedures used to examine linear, causal relationships between variables. It is defined as, "a statistical method for analyzing quantitative data that yields empirical estimates of the effects of variables in a *hypothesized causal system*" (Bohrnstedt & Knoke, 1994, p. 414). It shares many features of multiple linear regression analysis (e.g., linearity in causal relationships). Path modeling is an extension of multiple regression analysis in that it allows us to more fully determine how IVs are related to the DV in a model. To understand what we mean here, we return to our discussion on methods of data entry (i.e., simultaneous, sequential, and statistical) in multiple regression in Chapter 5. Recall that the results of simultaneous multiple regression analysis provide information about the direct effects of the IVs on the DV whereas the results of sequential multiple regression analysis provide information about the total effects of the IVs on the DV. In path analysis, we can investigate direct effects, total effects, and indirect effects, which can lead to stronger claims about the causal relationship between the variables. As we discussed in the Introduction, four conditions must be met in order to establish a causal relationship between two variables—association, temporal ordering (i.e., time precedence), correct effect priority, and nonspuriousness (i.e., no other plausible explanations of the relationship between the presumed cause and the presumed effect) (Kline, 2011). Path analytic techniques build on multiple regression analysis to lend more explicit support for a causal relationship between variables, particularly with regard to time precedence and nonspuriousness.

Path analysis is part of the structural equation modeling family. While **structural equation modeling** (SEM) combines measurement and structural models to analyze both latent and observed variables, path analysis focuses strictly on structural models with observed variables. However, the basic principles of path analysis provide the foundation for analyzing more complicated models in the SEM family, including measurement models and those with latent variables.

Structure of the Chapter

The first part of this chapter provides a detailed explanation of path analysis, including correlation; path diagrams, equations, and coefficients; path decompositions; assumptions; model identification; and model estimation. We refer to a study by Stewart (2006) described in the following paragraph, as a means of illustration. This section concludes with a description of examples from social work research that have used path analytic methods. We then provide an overview of structural equation modeling. At the end of the chapter, instructions for conducting path analysis in AMOS (SPSS Inc., 2011) are provided, using data from the National Educational Longitudinal Study of 1988 (NELS: 88).

Description of the Illustration

A study by Stewart (2006) is used to introduce and illustrate the principles and concepts of path analysis throughout the chapter. The data used for this study come from NELS: 88 data. Stewart (2006) examined the influence of individual and family factors on academic achievement among African American youth. Figure 7.1 provides a path diagram of her model. Stewart hypothesized that, among other factors, caregiver involvement in youths' schools and academic achievement in the eighth grade would have a direct effect on achievement in 12th grade. She also hypothesized that these factors would have indirect effects on 12th grade

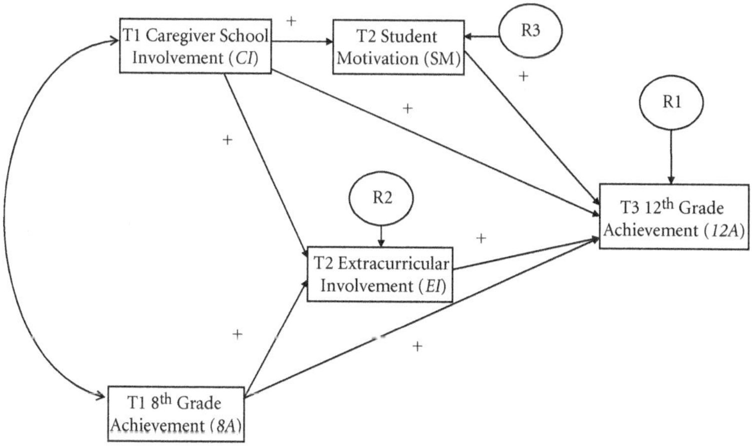

Figure 7.1 Proposed Path Diagram for Academic Success among African American Youth Model. Adapted from "Family- and Individual-Level Predictors of Academic Success for African American Students," by E. B. Stewart, 2006, Journal of Black Studies, 36, p. 605. Copyright 2006 by Sage Publications.

achievement through the youths' involvement in extracurricular activities during the 10th grade. Finally, caregiver school involvement in the eighth is also hypothesized to positively influence 12th grade achievement indirectly through student motivation in the 10th grade. Thus, Stewart posits that extracurricular activity involvement is both a cause of 12th grade achievement and an effect of caregiver–school involvement and academic achievement in the eighth grade. She also posits that student motivation in the 10th grade is both a cause of 12th grade achievement and an effect of caregiver involvement in the eighth grade.

The signs near each path indicate the hypothesized direction of the proposed relationships. In this case, all relationships are expected to be positive. That is, increased caregiver involvement and academic achievement during the eighth grade, and extracurricular involvement and student motivation in the 10th grade will lead directly to higher 12th grade achievement. Moreover, increased caregiver involvement and achievement during the eighth grade is hypothesized to lead to extracurricular involvement in the 10th grade, which will then lead to higher achievement in the 12th grade. Finally, increased caregiver involvement is expected to result in increased student motivation in the 10th grade, which will then also lead to higher achievement in the 12th grade.

The curved double-headed arrow connecting caregiver–school involvement and achievement in the eighth grade indicates that Stewart expects these factors to co-vary, but she does not offer any hypotheses about the causal relationship between them. Also, each DV (10th grade extracurricular involvement, 10th grade student motivation, and 12th grade achievement) has an uncorrelated error or residual term connected to it (i.e., *R*1, *R*2, and *R*3) denoted as a circle with a single-headed arrow pointing to its respective variable. These residual terms represent effects from other unknown causes of 10th grade extracurricular activity involvement, 10th grade student motivation, and 12th grade achievement beyond those included in the model and shown in the diagram.

PATH ANALYSIS: A DETAILED DESCRIPTION

Correlation as a Basic Statistic in Path Analysis

As with multiple linear regression analysis, **correlation** plays a major role in path analysis. In both techniques, we estimate the correlation between a set of IVs and a DV, measured on an interval or ratio scale, while controlling for other IVs. However, in path analysis we build on the process of controlling for other IVs by **decomposing** the correlation between each IV and DV into its component parts—direct effect, indirect effect, and/or spurious correlation.

Let's examine direct effects, indirect effects, and spurious correlations in more detail. A **direct effect** is represented as a path between two variables, *without* an intervening variable. Figure 7.1 provides several examples of direct effects. For instance, the path from T1 Caregiver School Involvement to T2 Student Motivation represents a proposed direct effect. An **indirect effect** is represented as a path between two variables, *with* an intervening variable between the other two variables. In Figure 7.1, the path from T1 Caregiver School Involvement to T2 Student Motivation to T3 12th Grade Achievement represents a proposed indirect effect of T2 Student Motivation (the intervening variable) in predicting the impact of T1 Caregiver School Involvement on T3 12th Grade Achievement. A **spurious correlation** occurs when the relationship between an IV and a DV is caused by another IV. For example, if the relationship between T1 Caregiver School Involvement and T2 Student Motivation is actually caused by Parental Educational Aspirations, then the correlation between T1 Caregiver School Involvement and T2 Student Motivation would be spurious. When a spurious correlation exists, then we have not met the condition of non-spuriousness as a part of the requirements for establishing causality.

Definitions of Other Important Terms

In previous chapters we introduced many terms that are used in linear regression analysis and ANOVA. Some of these terms are also used in path analysis. However, other terms that we have not encountered yet are associated with path analysis. We have already introduced you to five terms—path analysis, nonspuriousness, direct effects, indirect effects, and spurious correlation. In this section we define 10 additional terms and use Figure 7.1 to provide examples. All terms are defined in the context of presumed causality in path analysis. Five of the terms are variables defined from a causality framework; the other five terms are components of path analysis.

Thus far, we used the term **independent variable** to represent factors that are hypothesized to be associated with an outcome, and the term **dependent variable** to represent the outcome. These terms are not typically used in path analysis. Instead, we use the term **exogenous variable** to refer to a variable that we hypothesize to *predict* an outcome. Exogenous variables are predetermined and, from a temporal-ordering perspective, occur before the outcome. The symbol *X* is used to represent an exogenous variable. Figure 7.1 includes two exogenous variables—T1 Caregiver School Involvement and T1 8th Grade Achievement. The outcome of interest, or what we intend to predict, is referred to as an **endogenous variable.** The symbol, *Y* is typically used to denote an endogenous variable. Three endogenous variables are included in Figure 7.1 T2 Student Motivation, T2 Extracurricular Involvement, and T3 12th Grade Achievement.

Relative to multiple regression analysis, one difference is that path analysis usually involves models with more than one outcome of interest. Another difference is that some endogenous variables can have dual roles such that they can be modeled as both predictors and outcomes. In Figure 7.1, T2 Student Motivation and T2 Extracurricular Involvement are examples of endogenous variables with dual roles. These types of variables are referred to as **intervening variables**. As Figure 7.1 shows, T2 Student Motivation and T2 Extracurricular Involvement intervene between the two exogenous variables (i.e., T1 Caregiver School

Involvement and T1 8th Grade Achievement) and the final endogenous variable, T3 12th Grade Achievement. Intervening variables are also called **mediating variables.**

The fifth variable term is **residual variable.** In path analysis, residual variables are similar to residual or error in linear regression analysis in that they represent the unexplained portion of the $X \rightarrow Y$ path (i.e., $1 - R^2$). Sometimes residual variables are referred to as disturbances or disturbance terms (e.g., Kline, 2011). As illustrated in Figure 7.1, each endogenous variable (i.e., T2 Student Motivation, T2 Extracurricular Involvement, and T3 12th Grade Achievement) has an associated residual variable (i.e., *R*1, *R*2, and *R*3). A residual variable represents other causes of the associated endogenous variable that are not included in the model. For example, the residual variable *R*2 represents all causes of T2 Student Motivation, in addition to T1 Caregiver Involvement. It is the explicit focus on cause that differentiates residuals in path analysis from residuals in linear regression analysis.

The other five terms represent components of path analysis. A **path model** is a causal representation of relationships among variables. A path model is **recursive** (versus nonrecursive), which means that all causal relationships are assumed to occur in one direction (represented with arrows pointing in one direction—left to right). Nonrecursive path models hypothesize reciprocal relationships between variables (with arrows going in both directions). A path model can be depicted using symbols in the form of an equation or as a diagram. A **path diagram** is a pictorial or visual display of cause and effect relationships among variables. A **path coefficient,** denoted as *P*, is the numerical estimate of the causal relationship between two variables. The path coefficient represents the amount of expected change in the variable representing the effect due to a one unit change in the variable representing the cause. **Path decompositions** involve determining and separating out the types of correlation between two variables. This might include direct effects, indirect effects, and/or spurious correlations. In the following section, we use these terms to explain the path model.

Components of a Path Model

Path Diagrams

A path diagram provides a picture of the path model. It is the graphic equivalent to describing the model with a set of equations, which we will cover in a subsequent section. Embedded in the diagram are the four causal assumptions—association, temporal ordering, correct effect priority, and nonspuriousness—that we hypothesize about the cause and effect relationships between the variables.

Path diagrams are made up of four geometric symbols—squares or rectangles, lines with single-headed arrows, lines with double-headed arrows, and circles. The squares or rectangles are used to represent the measured variables, be they exogenous or endogenous. Figure 7.1 includes five rectangles representing five measured variables—T1 Caregiver School Involvement, T1 8th Grade Achievement, T2 Student Motivation, T2 Extracurricular Involvement, and T3 12th Grade Achievement.

The lines indicate relationships between variables. These lines represent the required condition of association in establishing causality. Lines with single-headed arrows (→) are used to represent the hypothesized impact of one variable (the cause) on another variable (the effect). These lines are the paths or relationships that will be tested in the analysis. Figure 7.1 includes seven single-headed arrow lines or paths connecting hypothesized causes (e.g., T2 Student Motivation) to their respective effects (e.g., T3 12th Grade Achievement).

Note that the single-headed arrow lines in Figure 7.1 have plus (+) or minus (−) signs above them. These signs depict the hypothesized direction of the causal relationship between the two variables. Thus, a path diagram shows that two variables are hypothesized to be associated with one another *and* what is expected in terms of the nature of their association. A plus sign indicates a positive association, such that the stronger the cause, the stronger the effect. A negative sign indicates an inverse association, such that the stronger the cause, the weaker the effect.

Lines with double-headed arrows (↔) are used to represent covariances or correlations between two variables. These relationships are assumed to already exist and are not tested in the analysis. Figure 7.1 has one double-headed arrow line, depicting the correlation between T1 Caregiver School Involvement and T1 8th Grade Achievement. Thus, Stewart (2006) posits that T1 Caregiver School Involvement and T1 8th Grade Achievement are correlated with one another.

The circles represent **residual variables**. Every endogenous variable has a residual variable connected to it. Recall that residual variables represent all of the other causes of the endogenous variable, beyond what is included in the model. As shown in Figure 7.1, a residual variable is connected to each endogenous variable. *R*1 represents all causes of 12th grade achievement, beyond T1 Caregiver School Involvement, T1 8th Grade Achievement, T2 Student Motivation, and T2 Extracurricular Involvement. *R*2 represents all causes of T2 Extracurricular Involvement, beyond T1 Caregiver School Involvement and T1 8th Grade Achievement. Finally, *R*3 represents all causes of T2 Student Motivation, beyond T1 Caregiver School Involvement.

Temporal ordering—another condition in establishing causality—is also depicted in a path diagram. By convention, temporal order is organized from left to right. This means that variables on the left side of a diagram are hypothesized to occur prior in time to variables that are toward the right side of the diagram. The Figure 7.1 diagram is based on three points in time—T1, T2, and T3. Two variables are measured at time 1 (i.e., T1 Caregiver School Involvement and T1 8th Grade Achievement). Two variables are measured at time 2 (i.e., T2 Student Motivation and T2 Extracurricular Involvement). The final outcome variable (i.e., T3 12th Grade Achievement) is measured at time 3.

Finally, the path diagram also shows the presumed effect priority, which is the direction of the causal influence from one variable to another. In our example, the direction of influence is straightforward due to the time dimension. The diagram reflects Stewarts' beliefs that factors that occur in the eighth grade predict factors that occur subsequently, in the 10th and 12th grades, and factors that occur during the 10th grade predict something that occurs in the 12th grade.

Bohrnstedt and Knoke (1994, p. 411) present six rules that guide the construction of path diagrams, as follows:

1. Variable names are represented by either short key words or letters.
2. Variables placed to the left in a diagram are assumed to be causally prior to those on the right.
3. Causal relationships between variables are represented by single-headed arrows.
4. Variables assumed to be correlated but not causally related are linked by a curved, double-headed arrow.
5. Variables assumed to be correlated but not causally related should be at the same point on the horizontal axis of the causal diagram.

6. The causal effect presumed between two variables is indicated by placing "+" or "–" signs along the causal arrows to show how increases or decreases in one variable affect the other.

Types of Relationships between Variables

Figure 7.2 shows diagrams of three types of cause and effect relationships—direct, mediating, and moderating, adapted from the diagram provided by Bohrnstedt and Knoke (1994, p. 412). Mediating and moderating cause and effect relationships are also referred to as indirect effects. Path analysis is used to test two types of cause–effect relationships—direct and mediating (i.e., intervening). The first diagram represents a direct cause–effect relationship between *X* (the IV) and *Y* (the DV). As you can see, Figure 7.1 proposes seven direct cause–effect relationships (i.e., T1 Caregiver School Involvement→T2 Extracurricular Involvement, T1 Caregiver School Involvement→T2 Student Motivation, T1 Caregiver School Involvement→T3 12th Grade Achievement, T1 8th Grade Achievement→T2 Extracurricular Involvement, T1 8th Grade Achievement→12th Grade Achievement, T2 Student Motivation→T3 12th Grade Achievement, and T2 Extracurricular Involvement→T3 12th Grade Achievement).

The second diagram in Figure 7.2 shows a mediating cause–effect relationship, with the addition of variable *Z*, occurring in a time-ordered manner between *X* and *Y*. In this case, *X* affects *Y* indirectly through *Z*. In other words, *Z* mediates the relationship between *X* and *Y*. The variable *Z* is also referred to as an intervening variable. Figure 7.1 provides two examples of mediating cause–effect relationships. First, T1 Caregiver School Involvement is hypothesized to affect T3 12th Grade Achievement indirectly through T2 Extracurricular Involvement. In other words,

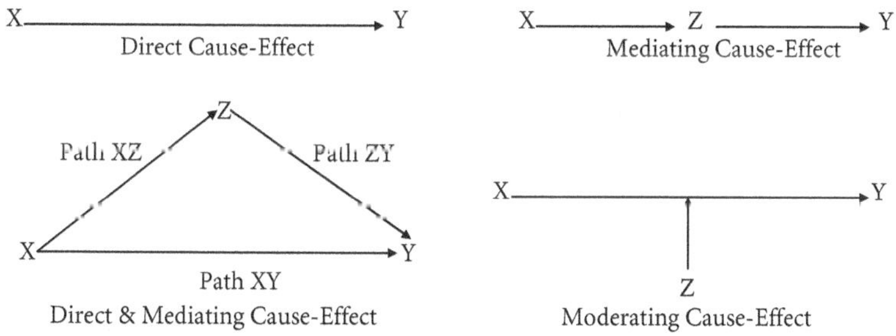

Figure 7.2 Types of Cause and Effect Relationships

extracurricular involvement is expected to intervene between or mediate the relationship between caregiver involvement and 12th grade achievement. As a second example, Figure 7.1 shows a hypothesized mediating cause–effect relationship between T1 8th Grade Achievement→T2 Extracurricular Involvement→T3 12th Grade Achievement, such that eighth grade achievement is hypothesized to affect 12th grade achievement indirectly through 10th grade extracurricular involvement.

The third diagram in Figure 7.2, the direct and mediating cause–effect relationship, depicts *X* as being both directly and indirectly related to *Y*. The direct effects on *Y* are represented as Path *XY* and Path *ZY*. The indirect effect of *X* on *Y* is represented as Path *XZ* and Path *ZY*. An example of direct and mediating cause–effect relationship is shown in Figure 7.1, whereby T1 8th Grade Achievement (*X*) affects T3 12th Grade Achievement (*Y*) directly as well as indirectly through T2 Extracurricular Involvement (*Z*).

We can distinguish between these mediating effects (commonly tested in path analysis), and moderating effects based on the interactions in two-way ANOVA models. Recall that in two-way ANOVA we are interested in determining whether one IV affects the relationship between another IV and the DV. In other words, we examine whether the impact of an IV on the DV is conditioned by another IV. This is displayed in the fourth diagram in Figure 7.2. Recall the Family Connections evaluation (Lindsey et al., 2010), described in Chapter 6. If the researchers were interested in examining whether study conditions and family structure interacted with one another to affect the outcome, they could test the interaction effect of the study condition and family structure on group means. This is referred to as a moderating effect because one IV (in this case, family structure) is hypothesized to alter the strength of the relationship between another IV (e.g., participation in Family Connections) and the DV (e.g., improved family functioning).

For further information on mediating and moderating variables, a widely cited source is Baron and Kinney (1986).

Path Coefficients

A **path coefficient** is an estimate of the direct, causal relationship between a causal variable (the IV) and an effect variable (the endogenous or outcome variable), controlling for other IVs in the sequence. Each line with a single-headed arrow in the path model has an associated estimated path coefficient. Thus, the path-analytic method produces as many estimated path coefficients as there are lines with single-headed arrows. A path coefficient is represented by the symbol, ρ, with two subscripts. The first subscript indicates the effect; the second subscript indicates the cause. For example, the notation for the estimated path coefficient reflecting the causal relationship between T1 Caregiver School Involvement→T2 Extracurricular Involvement is ρ_{EICI}, where ρ is the estimated coefficient; $_{EI}$ represents the effect variable, T2 Extracurricular Involvement; and $_{CI}$ represents the causal variable, T1 Caregiver School Involvement.

Recall our discussion of estimated regression coefficients in Chapter 5. Estimated path coefficients are simply regression coefficients that are standardized. Standardizing the estimates provides a common unit by which we can compare the impact of the estimates on the outcomes. Similar to the unstandardized estimates, standardized estimates are also considered effect size measures, and CIs can be calculated to determine their precision relative to the population value. The common unit also allows us to decompose (i.e., separate) correlations among variables. We will go over path decomposition in more detail in a subsequent section. As with estimated regression coefficients, path coefficients indicate the direction and relative magnitude of the effects of causal IVs on outcome variables, controlling for other variables in the sequence. They are interpreted similarly to estimated regression coefficients. That is, a path coefficient indicates the amount of expected change in the outcome or endogenous variable as a result of a one unit change in the causal variable, controlling for other variables in the sequence.

Path coefficients are estimated by regressing each endogenous variable on the IV in the equation. For example, to estimate ρ_{31} in Figure 7.3, we regress X_3 on X_1. Recall our discussion of the slope, b (i.e., beta) in Chapter 5. The standardized version of b is β. The path coefficients in path analysis are simply the standardized betas or β, which are correlation coefficients. Thus, in Figure 7.3, $\rho_{31} = \beta_{31;}$ $\rho_{32} = \beta_{32}$; $\rho_{41} = \beta_{41}$; $\rho_{42} = \beta_{42}$; and $\rho_{43} = \beta_{43}$.

Path Equations

Each single-headed arrow line or path in a path diagram is represented by an equation that depicts the relationship between the connecting variables. We use Figure 7.3 to illustrate how to determine the components of these equations. Figure 7.3 is simply a diagram of a generic path model.

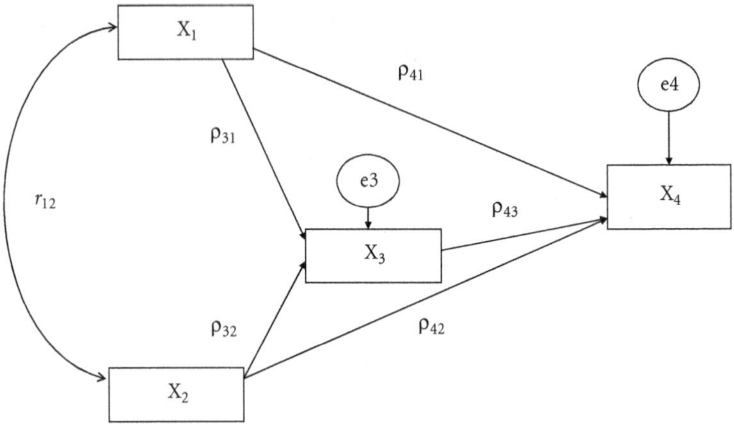

Figure 7.3 Example of a Basic Path Diagram

The exogenous variables are X_1 and X_2, the endogenous variables are X_3 and X_4, and the residual variables are *e*3 and *e*4. Recall that the residual variables represent the causal factors that are not in the model. As Figure 7.3 specifies, X_1 and X_2 are hypothesized to cause X_3. X_3 is also affected by variables not in the model, represented by *e*3. In this diagram, the causes of X_4 are more complicated, relative to the causes of X_3. X_4 is hypothesized to be caused directly by X_1, X_2, and X_3, and indirectly via the effects of X_1 and X_2 through X_3. The effect of other unspecified variables on X_4 is represented in *e*4.

Next to each path is a symbol, ρ, that represents its path coefficient. Following each ρ are two subscripts indicating the connecting effect variable and the causal variable, in that order. For example, the symbol ρ_{41} represents the estimated path coefficient of $X_1 \rightarrow X_4$, whereby X_1 is the causal variable and X_4 is the effect variable. The correlation between the two exogenous variables (which is not being examined in this analysis) is represented with the symbol, r_{12}.

We are now ready to develop the equations. The overall model has as many equations as the number of endogenous variables. For example, Figure 7.3 has two endogenous variables—X_3 and X_4. That means the overall model can be represented in two equations. Each equation includes as many terms as there are arrows leading to the particular endogenous variable. X_3 has three arrows pointing to it, each representing a causal factor X_1, X_2, and *e*. Thus, the equation for X_3 includes three terms. The equation is: $X_3 = \rho_{31}X_1 + \rho_{32}X_2 + e_3$. The second equation predicts X_4, which has four arrows pointing to it. This equation includes four terms, as follows: $X_4 = \rho_{41}X_1 + \rho_{42}X_2 + \rho_{43}X_3 + e_4$. Note that, in this model we hypothesize that X_4 is also affected by X_1 and X_2 indirectly through X_3. To capture the full nature of the relationship between X_3 and X_4 (both the direct and indirect effects), it is necessary to decompose this relationship. We will explain how to do this in the following section on path decompositions.

Path Decompositions

One of the strengths of path analysis is that it allows us to separate out the components of the path coefficient (i.e., correlation between two variables), also called path decomposition. By decomposing these path correlations, we gain a more complete understanding of the causal relationships. As indicated previously, we decompose only those path correlations that represent both direct and indirect (i.e., mediated) effects, as is shown in the path to X_4 in Figure 7.3.

As explained by Olobatuyi (2006), there are *up to* four components that make up a path correlation, depending on the type of relationship between the two variables it represents—direct effect, indirect effect, unanalyzed correlation, and/or effects from factors not in the model, also referred to as spurious effects. Particular correlations are made up of a mix of these components, depending on the hypothesized relationships. For instance, unanalyzed correlations are only associated with correlated exogenous variables (e.g., the correlation between X_1 and X_2 [r_{12}] in Figure 7.3). Furthermore, effects from factors not in the model (spurious effects [e.g., *e*3 and *e*4 in Figure 7.3]) are only associated with endogenous variables (e.g., X_3 and X_4 in Figure 7.3). Finally, some variables may only be *indirectly* related to other variables, with no direct effects.

Typically the focus is on decomposing correlations that include direct and indirect effects. However, to get a broad understanding of path decomposition, Table 7.1 (Olobatuyi, 2006, p. 111) provides the components of all path correlations in Figure 7.3. The first column lists the notation for the total association of each path in the model, represented as *r*. Note here that the subscript notations differ from the path-coefficient notations. That is, subscripts for *r* occur in chronological order. The second column lists the component representative

Table 7.1 Path Decomposition for the Diagram in Figure 7.3

	Path Decompositions			
Total Association	*Direct Effect*	*Indirect Effect*	*Unanalyzed correlation*	*Spurious Effects*
$r_{13} =$	P_{31}	0	$\rho_{32}r_{12}$	0
$r_{23} =$	P_{32}	0	$\rho_{31}r_{12}$	0
$r_{14} =$	P_{41}	$\rho_{43}\rho_{31}$	$\rho_{42}r_{12} + \rho_{43}\rho_{32}r_{12}$	0
$r_{24} =$	P_{42}	$\rho_{43}\rho_{32}$	$\rho_{32}r_{12} + \rho_{43}\rho_{31}r_{12}$	0
$r_{34} =$	P_{43}	0	0	$\rho_{41}\rho_{31} + \rho_{41}\rho_{32}r_{12} + \rho_{42}\rho_{31}r_{12} + \rho_{42}\rho_{32}$

Note. Adapted from A User's Guide to Path Analysis, by M. E. Olobatuyi, 2006, p. 111, University Press of America. Copyright 2006 by University Press of America.

of each path's direct effects. The third column lists the components that represent the indirect effects of the respective total correlation. The fourth column lists the components that represent unanalyzed correlation, and the final column indicates the components representing spurious effects. The information in Table 7.1 can then be used to determine the components of each path coefficient. For example, the correlation, r_{13}, is made up of a direct effect, ρ_{31}, and unanalyzed correlation, $\rho_{32}r_{12}$.

Path Analysis Assumptions and Requirements

Due to their similarity, path analysis and linear regression analysis share many of the same assumptions. This section provides a brief description of these assumptions and other model requirements for path analysis. For more details, refer to Chapter 5.

Assumptions about Variables

- The distribution of all variables approximates the shape of a normal curve.
- Variables are measured at the interval or ratio-level.
- Exogenous variables are measured without error.
- Exogenous variables are not correlated with variables not in the model.
- None of the causal variables are perfectly correlated with one another (i.e., no multicollinearity).

Assumptions about Relationships

- The relationship between the cause and effect variables is linear.
- The relationship between the cause and effect variables is additive. This means that there are no moderating (i.e., interaction) effects.
- Causation occurs in one direction. In other words, X is assumed to cause Y, rather than vise versa, and X and Y are not reciprocally related to one another.
- The path model has been accurately specified. That is, relevant variables have been included in the model and the causal ordering of variables is correct.

Assumptions about Residuals

- The mean value of the residuals for all values of X_i is zero.
- The variance of the residuals is constant for all values of X (i.e., homogeneity of variance).
- The distribution of the residuals should approximate the shape of a normal curve.
- The residuals of exogenous variables are independent of or uncorrelated with one another (i.e., lack autocorrelation).
- The residuals of each endogenous variable are not correlated with all other causes of that variable.

Other Model Requirements

- The minimum recommended sample size is 10 cases to every parameter in the model (Kline, 1998).
- The path model is **identified.** This means that a unique estimate can be determined for each parameter in the model. Model identification is discussed in the next section.

Model Identification

One of the requirements of path analysis is that there must be as many observations as there are parameters, such that the model is "just identified." Before we discuss what this means in terms of path analysis, it is useful to define **observations** and **parameters** in the context of path analysis and identified models. In this case, observations are *not* the number of cases in the study. Rather, they are the number of variances and *unique* co-variances among the observed variables. We can understand what is meant by "unique" co-variances by considering a variance–co-variance matrix. In a variance–co-variance matrix, the variances are provided along the diagonal. The co-variances between pairs of variables are provided twice in the matrix—both above and below the diagonal. In path models we include only one set of these co-variances.

The formula for determining the number of observations is: $v(v + 1)/2$, where v is the number of observed variables in the model. For example, the path diagramed in Figure 7.3 includes four observed variables (X_1, X_2, X_3, and X_4). To determine the number of observations in this model we replace "v" with "4" and solve the equation. Thus, the number of observations in this model is $4(4 + 1)/2 = 10$.

Model parameters are the estimates that are produced in the analysis. Parameters include all estimated path coefficients, the correlations among the exogenous variables, and the variances and co-variances of the exogenous variables, including the residuals. For example, in Figure 7.3 we estimate 10 parameters, including five path coefficients (i.e., ρ_{31}, ρ_{32}, ρ_{41}, ρ_{42}, and ρ_{43}); the two variances for X_1 and X_2, respectively; the co-variance between X_1 and X_2; and the two variances for each of the disturbances (i.e., e_3, and e_4). Thus, the diagram in Figure 7.3 is just identified because the number of observations and the number of parameters are equal.

Because path models are always recursive (i.e., the relationship between X and Y is unidirectional [$X \rightarrow Y$]) and the disturbances (residuals) are assumed to be independent (i.e., uncorrelated), it is impossible to specify a path model that is under identified (i.e., more parameters than observations) or over identified (i.e., more observations than parameters). Thus, if your path model is correctly specified, it is not necessary to be concerned about model identification in path analysis.

Estimating Path Models

Overview

Path models can be estimated using either standard regression-analysis methods or with a model fitting program, such as AMOS (SPSS Inc., 2011) or Mplus (Mplus). In multiple regression methods, one must conduct a separate analysis for each endogenous variable. Model-fitting programs estimate model parameters simultaneously. They also provide statistics that can be used to evaluate model fit; this information is not provided in multiple regression analysis procedures. We describe how to estimate path models using a model-fitting program (i.e., AMOS) rather than multiple regression analysis.

Three sets of parameters are estimated in path models—1) estimated path coefficients, representing direct and indirect effects, 2) correlations between exogenous variables, and 3) the variances and covariances of the disturbance terms. Their interpretations are straightforward. The interpretation of estimated path coefficients is similar to the way we interpret estimated standardized coefficients in linear regression analysis. The correlation between pairs of exogenous variables (i.e., Pearson *r*) is interpreted as the degree of association between them. Finally, estimates of disturbance term variances for endogenous variables indicate the portion of variation that is not explained by the variables in the model.

Maximum Likelihood Estimation

Model-fitting programs use **maximum-likelihood (ML) estimation** rather than the least squares method used in linear regression analysis. The ML estimation process is based on the assumption that the estimated parameters represent population values. Thus, the intent of ML estimation is to maximize the likelihood that the obtained sample estimates reflect the population parameters as closely as possible. In ML estimation, the parameter estimates are calculated simultaneously through an iterative process. What happens here is that the statistical program comes up with an initial set of estimates and then runs through multiple sets of calculations, seeking to determine the "best" fitting set of estimates. This iterative process continues until it converges (or not) on a solution. If iterative estimation is unsuccessful and fails to converge, this will be indicated as part of the output. In previous chapters, we identified several assumptions that are necessary in order to make inferences about our findings. The same assumptions are required in ML estimation.

Evaluating Model Fit

In path analysis with ML estimation, various tests are available to evaluate the fit of the model to the data. These tests vary with regard to which aspect of fit is being evaluated. Thus, it is common to use several statistical fit indices to evaluate model fit. We review two types of fit indices used to assess model fit in path analysis—absolute fit indices and incremental fit indices. For more information on fit indices, see Kline (2011).

Absolute fit indices. Absolute fit indices are used to compare the observed variances and covariances to the expected variances and covariances, or what Kenny (2011) refers to as "the perfect model." The larger the value of the statistical index, the further the observed model is from the perfect or expected fit. Thus, we seek absolute fit indices with relatively small values. Common absolute fit indices are the Chi Square test, the Goodness of Fit Index (GFI), the Standardized Root Mean Square Residual (SRMR), and the Root Mean Square Error of Approximation (RMSEA).

Comparative fit indices. Comparative fit indices—also referred to as incremental fit indices—are used to assess the improvement in fit of the observed model relative to the baseline or null model. Fit values range from 0 to 1.0. Larger values indicate better fit. In fact, values that are less than .90 are considered unacceptable. Common incremental fit indices are the Comparative Fit Index (CFI) (Bentler, 1990), the Normed Fit Index (NFI) (Bentler & Bonett, 1980), and the Non-Normed Fit Index (NNFI) (Bentler & Bonett).

Path Analysis in Social Work Research: Examples from the Literature

Path analysis has been used to inform important issues to social work researchers. For instance, Azuma and Chasnoff (1993) used path analysis techniques to examine the impact of *in utero* drug exposure on subsequent developmental outcomes for toddlers. The researchers used a prospective, longitudinal design to determine the direct effects of drug exposure (i.e., exogenous variable) on IQ (the outcome endogenous variable) and whether the effects of drug exposure on IQ were mediated by three other endogenous variables—head circumference, the home environment, and problematic behaviors (i.e., indirect effects). The researchers found that prenatal drug exposure had a direct effect on lower IQ scores for toddlers at age 3. Indirect effects were also reported through all three of the endogenous variables.

In another example, Ellickson and Hayes (1991) employed path analytic techniques to examine the direct effects of several cognitive (e.g., alcohol expectancies), social (e.g., peer alcohol use), and behavioral (e.g., academic performance) factors on the frequency of heavy alcohol use among seventh graders at two subsequent time points—3 and 12 months. The researchers also examined whether the frequency of heavy alcohol use at 3 months mediated relations between the exogenous variables (i.e., cognitive, social, and behavioral factors) and the frequency of heavy alcohol use at 12 months. Several variables (e.g., alcohol expectancies) had direct effects on heavy drinking at 12 months. Many of these relationships (e.g., peer alcohol use) were also mediated by heavy alcohol use at 3 months.

Finally, Kogan and others (2005) examined whether high school dropout status mediated the relationship between prior drug use and problem behavior on subsequent drug use among 1,762 African American youth. The data came from the 1990 and 1992 follow-ups of the NELS: 88. Path modeling indicated that marijuana use, cigarette use, and problem behavior at time 1 had direct effects on high school dropout at time 2 and subsequent marijuana and cigarette use at time 3. Results also showed that dropout status mediated the impact of the time 1 measures on drug use at time 3.

STRUCTURAL EQUATION MODELS WITH LATENT VARIABLES: AN INTRODUCTION

In the previous section we described path analysis models, which belong to the **structural equation modeling** (SEM) family. In this section we build on the path analysis presentation to present an overview of structural equation modeling with latent variables. Recall that, in path analysis, we assume that residuals are not correlated with one another and that causation occurs in a unidirectional fashion (i.e., one way from left to right), as reflected in a recursive model. SEM permits correlated error among residuals as well as reciprocal relationships between variables (i.e., nonrecursive models). In SEM, endogenous variables are referred to as **latent variables.** A latent variable cannot be directly observed, and, as a consequence, is measured indirectly through observed variables or indicators. By the way, the conventional symbol for a latent variable is a circle.

Examples of latent variables are attitudes toward homosexuality, perceptions of safety, and extroversion (vs. introversion). One well-known latent variable is self-esteem, measured using the Rosenberg Self-Esteem Scale (Rosenberg, 1965). The Rosenberg Self-Esteem Scale is made up of

10 items that describe aspects of self-esteem. Respondents are asked to select one of four choices (i.e., strongly agree, agree, disagree, or strongly disagree) that most closely indicates their views of each item. Responses to all items are then combined to determine a measure of self-esteem for each respondent.

SEM with latent variables involves a combination of factor-analytic and path analytic techniques, representing two models. One model is called the **measurement model** and is tested using factor analysis. The other model is called the **structural model** and is estimated using path analysis techniques. Let's focus first on the measurement model. The purpose of the measurement model is to confirm that hypothesized indicators (i.e., observed variables) represent latent variables (i.e., unobserved variables), albeit imperfectly. Through this process we attempt to operationalize concepts, such as self-esteem, that are difficult to measure because they are not easily observable. Applying the self-esteem example to an SEM framework, we would test the measurement model using factor-analytic techniques to determine the extent to which responses to the 10-item Rosenberg Self-Esteem Scale actually measure the latent or unobserved variable—self-esteem.

Once we have established relationships between observed and unobserved or latent variables, we then test the second component of SEM, the structural model. This is simply a test of the factors derived in the measurement model using the path analytic techniques we discussed previously. However, in this case, the endogenous variables are often latent or unobserved, and causal relationships may be reciprocal rather than just unidirectional. For those of you who are interested in learning more about structural equation modeling, we recommend Bowen and Guo (2012).

CONDUCTING PATH ANALYSIS IN AMOS: EXAMPLE

To summarize, path analysis is used to determine causal relationships between exogenous and endogenous variables. Endogenous variables must be measured on an interval or ratio-level. The following provides instructions for conducting path analysis in AMOS, using data from NELS: 88. A more detailed set of instructions is available on the companion website of this book.

Socioeconomic status (BYSES), self-concept (BYCNCPT1), and time spent on English homework each week (BYS79C) at time 1 are modeled as exogenous variables predicting time 1 standardized reading scores (BY2XRSTD); time 1 standardized reading scores (BY2XRSTD) are modeled as an endogenous variable predicting time 2 standardized reading scores (F12XRSTD). The data come from a random sample of 1,500 cases selected from NELS:88/92 dataset with no missing observations in any of the variables. The SPSS data set used for this example is PATH ANALYSIS.sav. IBM AMOS version 19 is used for this demonstration.

- Click IBM AMOS 19 and click Amos Graphics to open AMOS.
- To import the SPSS data file into AMOS, click "File."
- A drop down menu will appear. Move your cursor to "Data Files" and click to open the Data Files dialog box.
- Click "File Name." Locate the SPSS data file, click on it and click open. This will bring you back to the Data Files dialog box. The data file is under "File." Click "OK."
- Specify the model by drawing a diagram that includes all of the variables. The exogenous and endogenous variables are represented as rectangles. To draw the BYSES rectangle, click "Diagram" and then click "Draw Observed."
- Move the mouse to the location for the BYSES rectangle in the drawing area. Press the left mouse button to draw the rectangle. Release the mouse button to complete the rectangle. Follow the same steps to draw the rectangles of the other four variables.
- Once the rectangles have been drawn, click the rectangle icon to deactivate it.
- Draw ovals to represent the residual variables for the endogenous variables. Click "Diagram" and then click "Draw Unobserved."
- Move the mouse to the location for the first residual variable in the drawing area. Press the left mouse button to draw the first oval. Release the mouse button. Follow the same steps to draw the oval of the other endogenous variable.
- Once the ovals have been drawn, click the oval icon to deactivate it.
- Define the variables. Double click on the first rectangle that represents BYSES (socioeconomic status). The Object Properties dialog box appears. Click the "Text" tab and type BYSES in the Variable name field. Close the Object Property box by clicking

"X" at the top right corner. Repeat this with the other rectangles and ovals. ***Remember that the names given to the observed variables must match the variable names in the SPSS data file.***

- Draw paths connecting the observed variables using single-headed arrows. To draw a single-headed arrow, click on the single-headed arrow (i.e., Draw Paths) icon from the toolbar (just below the rectangle icon). Click and hold down the left mouse button from the right edge of BYSES to the left edge of BY2XRSTD rectangle. Clicking the BYSES rectangle will illuminate the edges of the rectangle in red. Release the mouse button when the BY2XRSTD rectangle edge is illuminated in green. Repeat this procedure for the remaining paths with single-headed arrows.
- When finished drawing the single headed arrow paths, click the single-headed arrow icon to deactivate it.
- Draw paths between the exogenous variables using double-headed arrows. To draw a double-headed arrow path, click on the double-headed arrow (i.e., Draw Covariances) icon. Click and hold down the left mouse button from the left edge of the BYCNCPT1 rectangle to the left edge of the BYSES rectangle. Repeat this procedure for the remaining paths with double-headed arrows.
- After drawing the double-headed arrow path, click on the double-headed arrow icon to deactivate it.
- Define the scale of the residual variables (V) and (B) by fixing the path coefficients from V to F12XRSTD and B to BY2XRSTD to 1. Double click on the path between V and F12XRSTD to open the Object Properties dialog box. Click on the Parameters tab and enter 1 in the Regression weight box. Click the X on the upper right hand corner of the dialog box to close it. Repeat these steps to define the scale of B.
- To perform the analyses, click "View" and then "Analysis Properties." Click the "Output" tab and check Minimization history, Standardized estimates, and Squared multiple correlations. Close the Analysis Properties window by clicking the X in the top right corner. Click "Analyze" and then "Calculate Estimates."
- The Save As dialog box will appear. Select the location where you want to save it, name the file, and click "Save." The saved file for

this example is available on the companion website. The name of the file is, "Path Analysis Tutorial."

- Once the file has been saved, the model estimates are calculated.
- To view the text output, click "View" and then "Text Output." Links to nine components of the output are provided in the left pane. To view a particular component, click on it. The details appear in the right pane. To exit the text output window, click on the X on the upper right hand corner of the dialog box.
- To view the graphics output, click the "View Output" icon next to the drawing area. The unstandardized estimates appear on the path diagram. To view the standardized estimates, click "Standardized estimates" in the third box below the output icon.
- NOTE: A copy of the output is available on the companion website. The file is named, "Path Analysis Output.pdf."

ADDITIONAL READING

Agresti, A. & Finlay, B. (2009). An introduction to advanced methodology. In A. Agresti & B. Finlay (Eds.), *Statistical methods for the social sciences* (4th ed.) (pp. 519–544). Upper Saddle River, NJ: Prentice Hall.

Keith, T. Z. (2006). *Multiple regression and beyond*. New York: Allyn & Bacon.

Knoke, D., Bohrnstedt, G. W., & Potter-Mee, A. (2002). Causal models and path analysis. In D. Knoke, G. W. Bohrnstedt, & A. Potter-Mee (Eds.), *Statistics for social data analysis* (4th ed.) (pp. 371–403). Itasca, IL: Wadsworth Publishers.

Olobatuyi, M. E. (2006). *A user's guide to path analysis*. Lanham, MD: University Press of America.

Appendix
Statistical Symbols

Symbol	*Meaning of Symbol*
μ	Population mean or average (Greek letter mu)
$\overline{X}$	Sample mean or average
σ	Population variance
σ^2	Population standard deviation
S^2	Sample variance
s	Sample standard deviation
χ^2	Chi-square
df	Degrees of freedom
r	Correlation coefficient
r^2	Coefficient of determination
α	Intercept (Greek letter alpha)
β	Slope
b_i	Unstandardized regression coefficient
β_i	Standardized regression coefficient
H_0	Null hypothesis
H_A	Alternative or research hypothesis
η^2	Effect size measure of the association between two variables (Greek letter eta)
ω^2	Effect size measure of the association between two variables with equal number of participants in each group (Greek letter omega)
ε	Residual error
ψ	Post-hoc comparisons (Greek letter psi)
α	Main effect for Factor A
β	Main effect for Factor B
$\alpha\beta$	Interaction effect for Factors A and B combined
ρ	Path coefficient

Glossary

Absolute fit index used to compare the observed variances and covariances to the expected variances and covariances in structural equation modeling

Adjusted mean used in ANCOVA; the change in the value of the group mean as a result of removing or controlling for the effects of the covariate

Analysis of covariance (ANCOVA) a statistical method used to determine whether means across groups are equal, after controlling for the effect of one or more covariate (i.e., variables hypothesized to explain some of the variance of the dependent variable, beyond that already hypothesized to be explained by the independent variable)

Analysis of variance (ANOVA) a statistical method used to determine whether the means across groups are equal

ANOVA summary table the format for presenting the sources of variation in ANOVA models

Association a relationship of association predicts that certain value categories of one nominal or ordinal variable will be associated or found more often with certain value categories of a second nominal or ordinal variable; one of the four conditions required for establishing a causal relationship between two variables

Attributes the possible values that can be assigned to a variable

Bar graph a graphical representation that displays frequencies or percentages of the attributes of a variable as either horizontal or vertical bars

Between-groups variability in ANOVA, one of two types of variance; differences of the group means relative to the combined mean for all groups

Between-groups sum of squares one of two components of total sums of squares in ANOVA; represents the variability or variation in scores across groups

Bimodal a distribution that has two values or attributes that occur with the most frequency

Bivariate statistical methods statistical methods used to analyze the relationship between two variables

Bivariate regression analysis linear-regression models that include only one independent variable; also referred to as **simple-regression analysis**

Causal a causal relationship predicts that the values of one variable (i.e., cause) results in the values of a second variable (i.e., effect)

Chi-square (χ^2) test a statistical test that measures the association between two variables measured at the nominal or ordinal level

Coefficient of determination indicates the proportion of the variation in one variable that is related to (or can potentially be explained by) the values of a second variable

Coefficient of multiple determination the proportion of variability in an analysis that is accounted for by the statistical model; also referred to as the **squared multiple correlation coefficient**

Column graph a graphical representation that displays frequencies or percentages of the attributes of a variable as columns

Comparative fit index used to assess the improvement in fit of the observed model relative to the base or null model in structural equation modeling; also referred to as **incremental fit index**

Conceptualization the process of defining a variable

Confidence interval a range of values around a given statistic (e.g., the mean) that are believed to contain with certain probability (e.g., 95%) the true value of that statistic (i.e., the population parameter)

Confounding variable an extraneous variable that correlates with both the independent and dependent variables and can lead to the incorrect conclusion that the independent and dependent variables are related

Constant a concept or characteristic in a research study that does not vary

Continuous variable a variable that can in theory take on an infinite value at any decimal point in a given interval

Control variable a variable whose potential effects on a dependent variable are held constant while observing an independent variable's influence on the dependent variable

Correct effect priority the direction of the causal influence has been accurately specified; one of the four conditions required for establishing causality between two variables

Correlation a relationship of correlation predicts that certain values of one variable will move either in the same direction (**positive correlation**) or in the opposite direction (**negative correlation**) as the values of a second variable

Correlation coefficient a numerical indicator of strength and direction of the relationship between the paired values of two variables

Covariate explanatory variable that is hypothesized to explain some of the variance of the dependent variable, beyond that of the independent variable

Cross-tabulation table a table displaying the categories or values of one variable as rows and the categories or values of another variable as columns with corresponding frequencies and percentages displayed in each cell

Decomposing in path analysis, dividing the correlation between an independent variable and a dependent variable into its component parts—direct effect, indirect effect, and/or spurious correlation

Degrees of freedom the number of values involved in the calculation of a statistic that are free to vary

Dependent-samples *t* test a statistical test that is used to determine if there is a statistical difference between the means of two related or paired samples; also referred to as **paired-samples *t* test**

Dependent variable the variable that the researcher believes will be influenced or affected by the independent variable; also referred to as outcome variable

Descriptive statistics methods for organizing and presenting quantitative data in ways that are concise and understandable

Dichotomous variable a variable that has only two possible values or attributes

Direct effect a path between two variables, without the influence of an intervening variable

Direction of influence one of the four conditions required for establishing causality between two variables; the presumed cause precedes the presumed effect

Directional hypothesis a hypothesis that predicts not only a significant relationship between variables, but also predicts the direction of the relationship; also referred to as **one-tailed hypothesis**

Discrete variable a variable that can only take on a finite value, typically reflected as a whole number

Dispersion the amount of variation that exists among the values of a distribution, also known as variability or spread

Effect priority the direction of the causal relationship between *X* and *Y* (e.g., *X* causes *Y* rather than the reverse) is correctly specified

Effect size strength and (typically) direction of the relationship between variables

Endogenous variable in a path model it is the variable to be predicted

Exogenous variable in a path model it is the variable that is hypothesized to predict an endogenous or outcome variable

Expected frequency the frequency that would be expected if the null hypothesis is true

Experiment-wise Type I error rate the Type I error rate that occurs as a result of conducting multiple tests in the same analysis; also referred to as **family-wise Type I error rate**

External validity the extent to which the results of a study can be generalized to other samples in the population

Factorial analysis of variance a specific type of ANOVA used to examine models with one dependent variable and mulitiple independent variables (i.e., factors)

Family-wise Type I error rate the Type I error rate that occurs as a result of conducting multiple tests in the same analysis; also referred to as **experiment-wise Type I error rate**

Forced entry multiple regression a method of data entry for conducting multiple linear regression analysis in which all the independent variables are entered into the equation at the same time; also referred to as **standard entry multiple regression**

Frequency distribution a summary of the frequencies for individual values or range of values for a single variable

Goodness of fit a comparison of how well a statistical model fits a set of actual data points

Heteroscedasticity having variances that are not equal; a violation of the assumption of homoscedasticity

Hierarchical multiple regression a method of data entry for conducting multiple linear regression analysis in which the independent variables are entered into the equation in a predetermined order, either individually or as groups; also referred to as **sequential multiple regression**

Histogram a graphical representation that displays frequencies or percentages of the attributes of a variable as a series of adjacent columns; generally used to represent continuous or numerical variables rather than categorical variables

Homogeneity of regression the assumption that a covariate relates to the dependent variable in exactly the same way for all groups of the independent variable

Homogeneity of variances having equal variances; also referred to as **homoscedasticity**

Homoscedasticity having equal variances; also referred to as **homogeneity of variances**

Incremental fit index used to assess the improvement in fit of the observed model relative to the base or null model; also referred to as **comparative fit index**

Independent-samples *t* test a statistical test that is used to determine if there is a statistical difference between the means of two unrelated or independent samples

Independent variable a variable that is predicted to explain or cause the values of another variable

Indirect effect the influence of one variable on another that occurs either fully or partially through a third variable; also referred to as mediating effect

Inferential statistics methods that use data observed in a sample to make predictions or hypotheses about data that have not been observed in the larger population

Interaction occurs when the effect of an independent variable on the dependent variable differs depending on the level of another variable

Interaction effect in two way ANOVA, it is the combined effect of two factors in explaining variation between group means; used to determine if the two factors operate independently of each other or if they work together to produce an additional effect on the dependent variable

Intercept the point at which a line crosses the vertical or *y*-axis of a graph

Internal validity the extent to which explanations for the study's findings can be attributed to the experiment, and alternative explanations have been ruled out

Interquartile range the range of values in a distribution that falls between the 25th percentile and the 75th percentile

Interval a level of measurement for numerical variables; equal distance between possible values; no absolute zero

Intervening variable a variable that comes between an independent and dependent variable; it is hypothesized to account for the effect of the independent variable on the dependent variable, either partially or fully; also referred to as a **mediating variable**

Kurtosis the degree to which the curve of a distribution is peaked or flat

Latent variable in structural-equation modeling, it is a variable that can not be directly observed and must be measured indirectly through observed indicators

Least squares estimation method the method for estimating the parameters in a statistical model by determining the best fitting straight line to solve a regression equation

Line chart a graphical representtation that displays frequencies or percentages of the attributes of a variable as a series of points connected by a line; generally used to represent continuous or numerical variables rather than categorical variables

Linear regression analysis a statistical procedure designed to determine the relationships between one or more independent variables and one dependent variable measured at the interval or ratio level

Marginal mean in two-way ANOVA, it is the mean for each level of each factor

Maximum-likelihood (ML) estimation a method of estimating the parameters in a statistical model in a way that most closely reflects the observed values

Mean the average of the values in a distribution calculated by adding all the values and dividing the sum by the number of values

Measurement systematic process that involves assigning labels (usually numbers) to characteristics of people, objects, or events using explicit and consistent rules so, ideally, the labels accurately represent the characteristic measured

Measurement model used in structural-equation modeling to confirm that hypothesized indicators (i.e., observed variables) represent latent variables (unobserved variables), albeit imperfectly

Measurement validity the degree to which accumulated evidence and theory support interpretations and uses of scores derived from a measure

Measures of central tendency estimations of the center or typical value of a data distribution

Measures of variability measures that indicate the amount of variation that exists among the values of a distribution, also known as dispersion or spread

Median the value in the middle of a distribution that has been ranked in order from the smallest to largest value

Mediator variable a variable that comes between an independent and dependent variable; it is hypothesized to account for the effect of the independent variable on the dependent variable, either partially or fully; also referred to as an **intervening variable**

Mode the value or attribute that occurs the most frequently in a distribution of values

Model sum of squares the sum of squares predicted by the regression model; represents the extent to which the prediction improved the fit of the model to the data, relative to using the mean alone

Moderating effect an effect of an independent variable on a dependent variable is conditioned on a third variable

Moderator variable a variable that alters the strength of the relationship between and independent and dependent variable

Multicollinearity the situation in which two independent variables in a multiple regression model are strongly correlated with one another

Multimodal a distribution that has three or more values or attributes that occur with the most frequency

Multiple correlation the correlation between the dependent variable and all of the independent variables in a multivariate regression analysis

Multivariate analysis of variance (MANOVA) a specific type of ANOVA used to examine models with more than one dependent variable and independent variable

Multiple linear regression models linear regression models that include more than one independent variable and one dependent variable measured at the interval or ratio level

Negative correlation predicts that certain values of one interval or ratio variable will move in the opposite direction as the values of a second interval or ratio variable

Nominal a level of measurement for categorical variables; attributes of a variable are mutually exclusive and exhaustive only

Nondirectional hypothesis a hypothesis that predicts a significant relationship between variables, but does not predict the direction of the relationship; also referred to as a **two-tailed hypothesis**

Nonorthogonal contrast in ANOVA, it is a test that compares the means between groups when the sources of variation overlap or are non-independent of one another

Nonparametric a statistical test that does not assume that the data have come from a probability distribution

Nonrandom sampling sampling techniques in which each element in the population has an unknown and usually unequal chance of being selected

Nonrecursive in structural equation modeling, it is a model in which reciprocal relationships between variables are hypothesized

Nonspuriousness the relationship between variables is not due to an outside variable; one of the conditions required for establishing causality between two variables

Normal distribution a distribution of data points that is symmetrical (i.e., bell-shaped) with most of the data points falling near the center of the distribution

Null hypothesis the hypothesis that presents the assumption or prediction that there is not a statistically significant relationship between variables

Observations in path analysis, the number of variances and unique co-variances among the observed variables

Observed frequency the actual frequencies that are observed in the data

Obscuring variable a type of moderating variable that affects the relationship between the independent and dependent variable such that, if the obscuring variable is not included in the analysis, the relationship between the independent and dependent variable is not statistically significant; also referred to as a **suppressor variable**

Omnibus F test one overall test that is conducted to determine statistical differences in means across multiple groups simultaneously; used to minimize the family-wise error rate

One-sample *t* test a statistical test that compares the mean of the dependent variable to the known or estimated mean of the population from which the sample was drawn

One-tailed hypothesis a hypothesis that predicts not only a significant relationship between variables, but also predicts the direction of the relationship; also referred to as **directional hypothesis**

One-way analysis of variance a statistical technique used to determine significant differences in means across groups

Operationalization the process of defining exactly how a variable will be measured

Ordinal a level of measurement for categorical variables; attributes are mutually exclusive and exhaustive, and also have an inherent order or ranking to them

Orthogonal contrast in ANOVA, it is a test used to compare the means between groups when the sources of variation are mutually exclusive or independent of one another

Outcome variable the variable that the researcher believes can be predicted, but not caused, by the values of the predictor variable

Outlier extreme cases within a distribution, or cases with values that have an unusually large impact on the results

***p* value** the probability of finding a relationship between variables in the sample as strong as the one observed when, in fact, there is not a real relationship between the variables in the population

Paired-samples *t* test a statistical test that is used to determine if there is a statistical difference between the means of two related or paired samples; also referred to as **dependent-samples *t* test**

Parameter a numeric value that summarizes a feature of a population

Parameters (in path analysis) estimates that are produced in path analysis, including all estimated path coefficients, correlations among the exogenous variables, and the variances and co-variances of the exogenous variables

Parametric test a statistical test that assumes that the data have come from a type of probability distribution; makes inferences about the parameters of the distribution

Partial correlation the correlation between one of the independent variables and the dependent variable in a multivariate regression analysis, controlling for the effects of all of the other independent variables on both variables

Partitioning the variance in ANCOVA, the sources of variation are separated into three components—between-groups adjusted, within-groups adjusted, and the covariate such that the variation attributable to between and within groups have been adjusted to account for the contribution of the covariate to the model

Path analysis a statistical procedure used to examine linear, causal relationships between variables

Path coefficient the numerical estimate of the causal relationship between a causal variable and an effect variable, controlling for other variables in the sequence

Path decomposition determining and separating out the types of correlation between two variables (e.g., direct effects, indirect effects, and/or spurious correlations)

Path diagram a pictorial or visual display of the cause and effect relationships among variables

Path model a causal representation of the relationships among variables that can be depicted as an equation or a diagram

Percentage distribution a summary of the percentages of the frequencies for individual values or range of values for a single variable

Percentile the value of a distribution below which a certain percentage of the values fall

Pie chart a graphical representation that displays frequencies of the attributes of a variable as percentages of a circle or pie

Planned contrast analyses conducted to test specific hypotheses about group mean differences that are developed *a priori* based on theory or previous research

Positive correlation predicts that certain values of one interval or ratio variable will move in the same direction as the values of a second interval or ratio variable

Post-hoc comparison analyses conducted to test specific hypotheses about differences in group means after the initial model has been run; often conducted in response to a significant omnibus *F* test; also referred to as **post-hoc contrast**

Power the power of a statistical test refers to the chance of avoiding a Type II error

Predictor variable the variable that the researcher believes can predict, but not cause, the values of a second variable, known as the outcome variable

Random assignment the process of randomly assigning participants to the different groups involved in a research study, such as the experimental and control groups

Random measurement error discrepancies between measured and actual ("true") values of a variable which are equally likely to be higher or lower than the actual values because they are caused by chance fluctuations in measurement; also known as "unsystematic" measurement errors

Random sampling sampling techniques in which each element in the population has an equal and known chance of being selected into the study

Range the distance that encompasses all of the values of a distribution, including the minimum and maximum values

Ratio a level of measurement for numerical variables such that there is equal distance between values with an absolute zero

Recursive a path model in which the influence of one variable on another is assumed to occur in one direction

Rejection level indicates how much difference between samples would be enough difference to be considered statisitically significant; established by the researcher prior to data collection; denoted by α (Greek alpha).

Reliability the degree of consistency of a method of measurement

Replication the process of repeating a research study two or more times using the same procedures but different research samples to see if the same results are found

Research hypothesis the hypothesis that states the assumption or prediction that there is a significant relationship between variables

Research sampling the process of obtaining participants for a research study

Residual the difference between the predicted data points and the actual data points, representing the error between the predicted and actual data in a regression analysis

Residual error represents the unexplained variance, or the portion of the dependent variable that is not explained by the independent variable

Residual plot a graphical representation (i.e., scatterplot) of the residuals versus the predicted outcome values; used in regression analysis to evaluate parametric test assumptions

Residual sum of squares the leftover sum of squares or the difference between the predicted data points and the actual observed data points; also referred to as **sum of squared residuals**

Residual variable represents all of the causes of an endogenous variable that are not included in the path analysis model

Sampling bias the intended or unintended systematic distortion of a sample that prevents the sample from being representative of the population from which it was drawn

Sampling distribution of the means the theoretical distribution that would be formed by drawing multiple samples from a population, calculating the means of each sample, and creating a sample or distribution with these means

Sampling distribution of the differences between means a theoretical distribution used to estimate whether two samples have been drawn from the same or different populations

Sampling error discrepancy between values in a population and values in a sample (i.e., discrepancy between parameters and statistics)

Scattergram a graph that shows the values of one variable along the x-axis and the values of a second variable along the y-axis with each corresponding pair plotted to suggest the relationship between the two variables

Semipartial correlation the correlation between one independent variable and the dependent variable in a multivariate regression analysis, controlling for the effects of the other independent variables on either the dependent variable or the independent variable, but not both

Sequential multiple regression a method of data entry for conducting multiple linear regression analysis in which the independent variables are entered into the equation in a predetermined order, either individually or as groups; also referred to as **hierarchical multiple regression**

Simple regression analysis linear regression model that includes only one independent variable; also referred to as **bivariate regression analysis**

Skew the degree to which the values of a distribution fall to one side of the mean or the other

Slope the gradient or steepness of a line

Spread the amount of variation that exists among the values of a distribution, also known as variability or dispersion

Spurious correlation exists when the relationship between an independent variable and a dependent variable is caused by another variable

Squared multiple correlation coefficient the proportion of variability in an analysis that is accounted for by the statistical model; also referred to as the **coefficient of multiple determination**

Standard deviation the positive square root of the variance; the average distance from the mean of each value in a distribution of values

Standard-entry multiple regression a method of data entry for conducting multiple linear regression analysis in which all the independent variables are entered into the equation at the same time; also referred to as **forced-entry multiple regression**

Standard error of the differences between means the standard deviation of the sampling distribution of the differences between means

Standard error of the mean the standard deviation of the sampling distribution of the means

Standard score the number of standard deviations a raw score falls above or below the mean; also referred to as a ***z* score**

Statistic a numeric value that summarizes a feature of a sample

Statistical multiple regression a method of data entry for conducting multiple linear regression analysis in which the order that the independent variables are entered into the equation is determined by statistics rather than by theory; also referred to as **stepwise multiple regression**

Stepwise multiple regression a method of data entry for conducting multiple linear regression analysis in which the order that the independent variables are entered into the equation is determined by statistics rather than by theory; also referred to as **statistical multiple regression**

Structural-equation modeling a family of statistical techniques used to estimate causal relations among variables

Structural model in structural equation modeling a model that represents hypothesized causal relationships between variables

Sum of squared residuals a measure of the leftover sum of squares or the difference between the predicted data points on the best fitting regression line and the actual observed data points; represents the errors in prediction; also referred to as **residual sum of squares**

Suppressor variable a type of moderating variable that affects the relationship between the independent and dependent variable such that, if the suppressor variable is not included in the analysis, the relationship between the independent and dependent variable is not statistically significant; also referred to as a **obscuring variable**

Systematic measurement error discrepancies between measured and actual ("true") values of a variable which are more likely to be higher or lower than the actual values of the variable; also known as "bias" in measurement.

Temporal ordering (i.e., precedence) in the order that something occurred in time; the presumed cause preceded the presumed effect; one of the conditions required for establishing causality between two variables

Total sum of squares a measure of the combined sum of squares predicted by the regression equation (i.e., model sum of squares) and the leftover or residual sum of squares

Trimmed mean the mean of a distribution of values calculated after disregarding a percentage of the lowest and highest values; serves to reduce the effects of outliers in the data

Two-tailed hypothesis a hypothesis that predicts a significant relationship between variables, but does not predict the direction of the relationship; also referred to as a **nondirectional hypothesis**

Two-way analysis of variance a type of ANOVA used to examine models with one dependent variable, measured at the interval or ratio level and two independent variables measured at the nominal levels

Type I error the chance of concluding that the relationship between variables is a real one (rejecting the null hypothesis) when, in fact, there is not a real relationship between variables in the population (equal to the p value)

Type II error the chance of concluding that there is no relationship between variables (failing to reject the null hypothesis) when, in fact, there is a real relationship between variables in the population

Univariate statistical methods statistical methods that organize and summarize data one variable at a time

Validity the accuracy of a method of measurement

Variable a concept or characteristic in a research study that varies

Variance the average squared distance between the mean and each value in a distribution of values

Weights values assigned to the coefficients of the equations that represent the contrasts

Within-groups variability (i.e., variance) in ANOVA, one of two types of variability; differences of the mean scores across participants in the same group, relative to the overall mean for that group

Within-groups sum of squares one of two components of total sums of squares in ANOVA; represents the variability or variation in scores within each group

***z* score** the number of standard deviations a raw score falls above or below the mean; also referred to as a **standard score**

References

Abu-Bader, S. H. (2006). *Using statistical methods in social work practice: A Complete SPSS Guide* Chicago, Ill: Lyceum Books.

Agresti, A., & Finlay, B. (2009). An introduction to advanced methodology. In A. Agresti & B. Finlay (Eds.), *Statistical methods for the social sciences* (4th ed.) (pp. 519–544). Upper Saddle River, NJ: Prentice Hall.

Alexander, K., Entwisle, D., & Horsey, C. (1997). From first grade forward: Early foundations of high school dropout. *Sociology of Education 70*(2), 87–107.

American Psychological Association. (2010). *Publication manual of the American Psychological Association* (6th ed.). Washington, DC: Author.

Arnold, E. M., Smith, T. E., Harrison, D. F., & Springer, D. W. (1999). The effects of an abstinence-based sex education program on middle school students' knowledge and beliefs. *Research on Social Work Practice, 9*(1), 10–24.

Auslander, W., Haire-Joshu, D., Houston, C., Williams, J. H., & Krebill, H. (2000). The short-term impact of a health promotion program for low-income African American women. *Research on Social Work Practice, 10*(1), 78–97.

Azuma, S. D., & Chasnoff, I. J. (1993). Outcome of children prenatally exposed to cocaine and other drugs: A path analysis of three-year data. *Pediatrics, 92*(3), 396–402.

Balfour, L., Tasca, G. A., Kowal, J., Corace, K., Cooper, C. L., Angel, J. B., Garber, G., MacPherson, P. A., & Cameron, D. W. (2007). Development and validation of the HIV Medication Readiness Scale. *Assessment, 14*, 408–416.

Baron, R. M., & Kenny, D. A. (1986). The moderator-mediator variable distinction in social psychological research: Conceptual, strategic, and statistical considerations *Journal of Personality and Social Psychology 51*(6), 1173–1182.

Bentler, P. M. (1990). Comparative fit indexes in structural models. *Psychological Bulletin*, 107, 238–246.

Bentler, P. M., & Bonett, D. G. (1980). Significance tests and goodness-of-fit in the analysis of covariance structures. *Psychological Bulletin, 88*, 588–600.

Bohrnstedt, G. W., & Knoke, D. (1994). *Statistics for social data analysis* (3rd ed.). Itasca, IL: F. E. Peacock.

Bowen, N. K., & Guo, S. (2012). *Structural equation modeling*. New York: Oxford University Press.

Chapman, J. F., Desai, R. A., Falzer, P. R., & Borum, R. (2006). Violence risk and race in a sample of youth in juvenile detention: The potential to reduce disproportionate minority confinement. *Youth Violence and Juvenile Justice, 4*, 170–184.

Clark, A., Skouteris, H., Wertheim, E. H., Paxton, S. J., & Milgrom, J. (2009). The relationship between depression and body dissatisfaction across pregnancy and the postpartum. *Journal of Health Psychology, 14*, 27–35.

Cohen, J. (1988). *Statistical power analysis for the behavioral sciences* (2nd ed.). Hillsdale, NJ: Lawrence Erlbaum Associates.

Cohen, J., Cohen, P., West, S. G., & Aiken, L. S. (2003). *Applied multiple regression/correlation analysis for the behavioral sciences* (3rd ed.). New York: Routedge.

Connors, N. A., Plybon, L. E., & Pegg, P. (2003). *Specific events drug and alcohol refusal self-efficacy scale*. Richmond, VA: Virginia Commonwealth University.

Cummings, G. (2012). *Understanding the new statistics: Effect sizes, confidence intervals, and meta-analysis*. New York: Routledge Publishing.

Edmond, T., Rubin, A., & Wambach, K. G. (1999). The effectiveness of EMDR with adult female survivors of childhood sexual abuse. *Social Work Research 23*(2), 103–116.

Ellickson, P. L., & Hays, R. D. (1991). Antecedents of drinking among young adolescents with different alcohol use histories. *Journal of Studies on Alcohol, 52*, 398–408.

Faulkner, C. A., & Faulkner, S. S. (2009). *Research methods for social workers: A practice-based approach*. Chicago, IL: Lyceum Books, Inc.

Field, A. (2005). *Discovering statistics using SPSS* (2nd ed.). Thousand Oaks, CA: Sage.

Field, A. (2009). *Discovering statistics using SPSS* (3rd ed.). Thousand Oaks, CA: Sage.

Foa, E., Cashman, L., Jaycox, L., & Perry, K. (1997). The validation of a self-report measure of PTSD: The Posttraumatic Diagnostic Scale. *Psychological Assessment, 9*, 445–451.

Garcia-Retamero, R., & Cokely, E. T. (2011). Effective communication of risks to young adults: Using message framing and visual aids to increase condom use and STD screening. *Journal of Experimental Psychology: Applied, 17*(3), 270–287.

Garner, D. M., Olmsted, M. P., Bohr, Y., & Garfinkel, P. E. (1982). The Eating Attitudes Test: Psychometric features and clinical correlates. *Psychological Medicine, 12*, 871–878.

Grissom, R. J., & Kim, J. J. (2012). *Effect Sizes for Research: Univariate and Multivariate Applications* (2nd ed.). New York: Routledge.

Horwitz-McCue, S., Balestracci, K. M. B., & Simms, M. D. (2001). Foster care placement improves children's functioning. *Archives of Pediatric and Adolescent Medicine 155*, 1255–1260.

Howell, D. C. (2010). *Statistical methods for psychology* (7th ed.). Belmont, CA: Cengage Wadsworth.

Hulsizer, M. R., & Woolf, L. M. (2009). *A guide to teaching statistics*. Malden, MA: Wiley-Blackwell.

Kagee, A. (2008). Symptoms of depression and anxiety among a sample of South African patients living with a chronic illness. *Journal of Health Psychology, 13*, 547–555.

Keith, T. Z. (2006). *Multiple regression and beyond*. Boston: Allyn & Bacon.

Kellner, R. (1986). The Brief Depression Rating Scale. In N. Sartorius & T. A. Ban (Eds.), *Assessment of depression* (pp. 179–183). New York: Springer-Verlag.

Kelly, V., Warner, K., Trahan, C., & Miscavage, K. (2009). The relationship among self-report and measured report of psychological abuse, and depression for a sample of women involved in intimate relationships with male partners. *The Family Journal: Counseling and Therapy for Couples and Families, 17*, 51–57.

Kenny, D. (2011, September 4). Measuring model fit [Web log post]. Retrieved from http://www.davidakenny.net/cm/fit.htm

Kim, J. A., Szatmari, P., Bryson, S. E., Streiner, D. L., & Wilson, F. J. (2000). The prevalence of anxiety and mood problems among children with autism and Asperger syndrome. *Autism, 4*, 117–132.

Kleinbaum, D. G., & Klein, M. (2010). *Logistic regression: A self-learning text*. New York: Springer Publishing.

Kleinbaum, D. G., Kupper, L. L., & Muller, K. E. (1988). *Applied regression analysis and other multivariate methods* (2nd ed.). Belmont, CA: Duxbury Press.

Kleinbaum, D. G., Kupper, L. L., Nizam, A., & Muller, K. E. (2007). *Applied regression analysis and other multivariable methods*. Belmont, CA: Thompson Publishing.

Kline, R. B. (1998). *Principles and practice of structural equation modeling* (2nd ed.). New York: The Guilford Press.

Kline, R. B. (2011). *Principles and practice of structural equation modeling* (3rd ed.). New York: The Guilford Press.

Knoke, D., Bohrnstedt, G. W., & Potter-Mee, A. (2002). Causal models and path analysis. In D. Knoke, G. W. Bohrnstedt, & A. Potter-Mee (Eds.), *Statistics for social data analysis* (4th ed.) (pp. 371–403). Itasca, IL: Wadsworth Publishers.

Kogan, S. M., Luo, Z., Brody, G. H., & McBride Murray, V. (2005). The influence of high school dropout on substance use among African American youth. *Journal of Ethnicity in Substance Abuse, 4*(1), 35–51.

Krysik, J. L., & Finn, J. (2010). *Research for effective social work practice* (2nd ed.). New York: McGraw Companies, Inc.

Leung, C., Tsang, S., Heung, K., & Yiu, I. (2009). Effectiveness of parent-child interaction therapy (PCIT) among Chinese families. *Research on Social Work Practice, 19*(3), 304–313.

Lev-Wiesel, R., & Liraz, R. (2007). Drawings vs. narratives: Drawing as a tool to encourage verbalization in children whose fathers are drug abusers. *Clinical Child Psychology and Psychiatry, 12*, 65–75.

Lindsey, M. A., Hayward, R. A., & DePanfilis, D. (2010). Gender differences in behavioral outcomes among children at risk of neglect: Findings from a family-focused prevention intervention. *Research on Social Work Practice 20*(6), 572–581.

Lomax, R. G. (1992). *Statistical concepts: A second course for education and behavioral sciences*. White Plains, NY: Longman Publishing Group.

Lomax, R. G. (2007). *Statistical concepts: A second course* (3rd ed.). New York Routledge Academic.

MacNeill, S., E., Lichtenberg, P. A., & LaBuda, J. (2000). Factors affecting return to living alone after medical rehabilitation: A cross-validation study. *Rehabilitation Psychology, 45*, 356–364.

Marlow, C. R. (2011). *Research methods for generalist social work* (5th ed.). Belmont, CA: Brooks/Cole.

McClatchey, I. S., Vonk, M. E., & Palardy, G. (2009). Efficacy of a camp-based intervention for childhood traumatic grief. *Research on Social Work Practice, 19*, 19–30.

M*plus* (Version 6.11) [Computer software]. Retrieved from www.statmodel.com/index.shtml. Los Angeles, CA: Muthen & Muthen.

Myers, L. L. (2010). Health risk behaviors among adolescents in the rural south: A comparison of race, gender, and age. *Journal of Human Behavior in the Social Environment, 20*, 1024–1037.

Nakonezny, P. A., Rodgers, J. L., & Nussbaum, J. F. (2003). The effect of later life parental divorce on adult-child/older-parent solidarity: A test of the buffering hypothesis. *Journal of Applied Social Psychology, 33*(6), 1153–1178.

Nasim, A., Belgrave, F. Z., Corona, R., & Townsend, T. G. (2009). Predictors of tobacco and alcohol refusal efficacy for urban and rural African American adolescents. *Journal of Child & Adolescent Substance Abuse, 18*, 221–242.

Ogles, B. M., Lunnen, K. M., Gillespie, D. K., & Trout, S. C. (1996). *Conceptualization and initial development of the Ohio scales.* Paper presented at the 8th annual research conference proceedings, A system of care for children's mental health: Expanding the research base Tampa, FL.

Olobatuyi, M. E. (2006). *A user's guide to path analysis.* New York: University Press of America.

Orme, J. G., & Combs-Orme, T. (2009). *Multiple regression with discrete dependent variables.* New York: Oxford University Press.

Painter, K. (2009). Multisystemtic therapy as community-based treatment for youth with severe emotional disturbance. *Research on Social Work Practice, 19*(3), 314–324.

Radloff, L. S. (1977). The CES-D scale: A self-report depression scale for research in the general population. *Applied Psychological Measurement, 1,* 385–401.

Rayle, A. D., Kulis, S., Okamoto, S. K., Tann, S. S., LeCroy, C. W., Dustman, P., & Burke, A. M. (2006). Who is offering and how often? Gender differences in drug offers among American Indian adolescents of the southwest. *Journal of Early Adolescence, 26,* 296–317.

Rosenberg, M. (1965). *Society and the adolescent self-image.* Princeton, NJ: Princeton University Press.

Rubin, A. (2010). *Statistics for evidence-based practice and evaluation* (2nd ed.). Belmont, CA: Thomson/Brooks Cole.

Rubin, A., & Babbie, E. (2008). *Research methods for social work* (6th ed.). Belmont, CA: Brooks Cole.

Russell, D. W. (1996). UCLA Loneliness Scale (version 3): Reliability, validity, factor structure. *Journal of Personality Assessment, 66,* 20–40.

Russell, D., Peplau, L. A., & Cutrona, C. E. (1980). The revised UCLA Loneliness Scale: Concurrent and discriminant validity evidence. *Journal of Personality and Social Psychology, 39,* 472–480.

Schiff, M., & Levit, S. (2010). Correlates of therapeutic alliance and treatment outcomes among Israeli female methadone patients. *Research on Social Work Practice, 20*(4), 380–390.

Shiovitz-Ezra, S., & Leitsch, S. A. (2010). The role of social relationships in predicting loneliness: The National Social Life, Health, and Aging Project. *Social Work Research, 34*(3), 157–167.

Silver, E., & Bauman, L. J. (2006). The association between sexual experience with attitudes, beliefs, and risk behaviors of inner-city adolescents. *Journal of Research on Adolescence 16*(1), 29–45.

Singh, N. N., Lancioni, G. E., Winton, A. S. W., Singh, J., Curtis, W. J., Wahler, R. G., & McAleavey, K. M. (2007). Mindful parenting decreases aggression and increases social behavior in children with developmental disabilities. *Behavior Modification, 31,* 749–771.

Sparrow, S. S., & Cicchetti, D. V. (1987). Adaptive behavior and the psychologically disturbed child. *Journal of Special Education, 21*, 89–100.

SPSS, Inc. (2011). *SPSS statistical software (Version 19.0) [Computer software]*. Chicago, IL: SPSS, Inc.

Stewart, E. (2006). Family- and individual-level predictors of academic success for African American students: A longitudinal path analysis utilizing national data *Journal of Black Studies 36*(4), 597–621.

Tabachnick, B. G., & Fidell, L. S. (1996). *Using multivariate statistics* (3rd ed.). New York: Allyn & Bacon.

Tabachnick, B. G., & Fidell, L. S. (2007). *Using multivariate statistics* (4th ed.). New York: Allyn & Bacon.

United States Department of Education. (2011). National Educational Longitudinal Study of 1988 (NELS: 88) [Data file]. Retrieved from http://nces.ed.gov/surveys/nels88/

Wang, X., Blomberg, T. G., & Li, S. D. (2005). Comparison of the educational deficiencies of delinquent and nondelinquent students. *Evaluation Review, 29*, 291–312.

Weinbach, R. W., & Grinnell, Jr., R. M. (2010). *Statistics for social workers*. Boston, MA: Allyn & Bacon.

Wetherill, G. B. (1986). *Regression analysis with applications*. London: Chapman & Hall.

Yegidis, B. L., Weinbach, R. W., & Myers, L. L. (2012). *Research methods for social-workers* (7th ed.). Boston: Allyn & Bacon.

Index

absolute fit indices, 180
analysis of covariance (ANCOVA), 7, 99, 133–134, 156–160
  adjusted mean, 157–158
  assumptions, 158–159
  between-groups adjusted, 158
  covariate, 7, 99, 134, 156–159
  examples in literature, 159–160
  within-groups adjusted, 158
analysis of variance (ANOVA), 6–7, 133–156, 160–162
  assumptions, 142–143
  examples in literature, one-way ANOVA, 149–150
  examples in literature, two way ANOVA, 155–156
  factorial ANOVA, 7, 134
  multivariate ANOVA (MANOVA), 7, 134
  one-way ANOVA, 7, 134–150
  repeated measures ANOVA, 134–135
  Summary Table, 140–141
  SPSS, calculate using, 160–162
  two-way ANOVA, 7, 134, 150–156
AMOS, 178, 182–185
  path analysis, 178, 182–185
association, 37–38
attribute, 13
autocorrelation, 123

bar graph, 20–23
basic statistics, 1–5, 69
bivariate statistics, 2
Bonferroni correction, 149
boxplot, 70

causal relationship, 5–6, 38–39, 103
categorical variable, 13
central tendency, measures of, 24–29
  mean, 24, 26–28
  median, 24–28
  mode, 24–25, 27–28
  selecting, 27–28
  SPSS, calculate using, 28–29
chi-square test of association, 76–85
  alternatives to, 84–85
  degrees of freedom, 80
  examples in the literature, 82–83
  SPSS, calculate using, 83–84
cluster analysis, 108
coefficient of determination, 105
coefficient of multiple determination, 119–120
column graph, 21–23
comparative fit indices, 180
confidence interval, 56–57, 65
confounding variable, 40–41, 86, 89
constant, 12

continuous variable, 13
contrasts, 143–149, 161–162
  nonorthogonal, 148
  orthogonal, 145–147
  planned, 144–148, 161
  post-hoc, 148–149, 161–162
control variable, 40
correlation, 38, 74–75, 99–108, 112, 120, 166
  alternatives to Pearson's *r*, 108
  bridge to advanced statistical tests, 108, 166
  causation, 103
  examples in the literature, 106
  negative, 38, 74–75, 101–102
  non-linear, 107–108
  partial, 120
  Pearson's product moment correlation coefficient (Pearson's *r*), 102–105, 112
  positive, 38, 74–75, 101–102
  semipartial, 120
  SPSS, using to calculation, 106–107
  spurious, 166
  strength of, 102–103
correlation coefficient, 102
covariate, 7, 99
cross-tabulation table, 72–74, 77

degrees of freedom, 80, 86, 89, 94, 121
  chi-square test of association, 80
  dependent samples *t*-test, 89
  F-test, 121
  independent samples *t*-test, 94
  one-sample *t*-test, 86
dependent samples *t*-test, 88–94
  degrees of freedom, 89
  examples in the literature, 92
  SPSS, calculate using, 93–94
dependent variable, 39
descriptive statistical methods, 2, 11
dichotomous variable, 13, 15
direct effect, 166
discrete variable, 13
dispersion, 29
Durban-Watson test, 123, 143

effect size, 65–67, 142
endogenous variable, 167, 169
exogenous variable, 167
expected frequency, 78–79
external validity, 50–51

F-test, 121, 139–140
  degrees of freedom, 121
factor analysis, 108
Fisher exact test, 85
Fisher's least significant difference, 149
frequency distributions, 18–20, 70
  cumulative frequency distributions, 20
  using grouped data, 19

G*Power, 126
goodness of fit, 119
graphical representation of distributions, 20–24
  bar graphs, 20–23
  column graph, 21–23
  histograms, 20, 23–24
  line graphs, 20, 24
  pie charts, 20–21

Hartley's F_{MAX} procedure, 71, 124
heteroscedasticity, 70–71, 123
hierarchical linear modeling, 71
histograms, 20, 23–24, 69
homogeneity of regression, 159
homogeneity of variance, 70–71, 123–124, 129–130
  SPSS, calculate using, 129–130
homoscedasticity, *see* homogeneity of variance
hypothesis testing, 58–67

independent samples *t*-test, 94–99, 137
  bridge to advanced statistical tests, 99, 137
  degrees of freedom, 94
  examples in the literature, 97–98
  SPSS, calculate using, 98–99
independent variable, 39
indirect effect, 166
inferential statistics, 3–5, 35, 55
  data assumptions, 4
interaction effect, 151–152, 155
intercept, 111,113–114
internal validity, 51, 54
intervening variable, 40, 167

Kendall's tau, 108
Krustal-Wallis test, 85, 149

kurtosis, 41–42, 49, 70
 leptokurtic, 41–42
 mesokurtic, 41–42
 platykurtic, 41–42

latent variable, 181–182
least-squares estimation, 113–116, 142, 153–154
levels of measurement, 15–17, 126–127
 interval, 16–17
 nominal, 15, 17
 ordinal, 16–17
 ratio, 16–17
Levene test for homogeneity of variance, 71, 123–124
line graph, 20
linear correlation, *see* correlation
linear regression analysis, 6–7, 109–130
 bivariate linear regression analysis, 110–112
 logistic regression, 110
 multiple linear regression analysis, 6–7, 110–113
 regression coefficients, 111
 simple regression model, *see* bivariate regression model
linearity assumption, 122
logistic regression analysis, 6
longitudinal panel study, 71

main effect, 151, 155
Mann-Whitney U, 85
marginal mean, 153
maximum-likelihood estimation, 179
mean sum of squares, 121
median, 24–29
 SPSS, calculate using, 28–29
mean, 24, 26–29
 SPSS, calculate using, 28–29
 trimmed mean, 28
 weighted mean, 26–27
measurement, 13–18, 52–53
 levels of measurement, 15–17
 measurement error, 14, 52–53
 random measurement error, 14, 53
 reliability of measurement, 14
 statistical levels of measurement, 17–18
 systematic measurement error, 14, 52–53
 validity of measurement, 14

Median test, 85
mediator variable, 39–40, 168
mode, 24–25, 27–29, 41
 bimodal, 25, 41
 multimodal, 25, 41
 SPSS, calculate using, 28–29
model estimation, 4–5, 136
model parameters, 136
moderator variable, 39–40
multicollinearity, 125
multiple analysis of variance (MANOVA), 99
multiple correlation coefficient (multiple *R*), 108, 119–120
multiple linear regression analysis, 6–7, 110–113, 127–130
 examples in literature, 128–129
 multiple correlation, 112
 partial correlation, 112–113
 semipartial correlation, 112–113
 sequential or hierarchical, 127–128
 simultaneous or forced entry, 127–128
 SPSS, calculate using, 129–130
 statistical or stepwise, 127–128
multivariate statistical analysis, 1–8

National Education Longitudinal Study of 1988, 9
nonparametric statistical test, 70–72
nonrandom sampling, 51–52
nonspuriousness, 164
normal distribution, 42–49, 70
null hypotheses, 36
obscuring variable, 40
observed frequency, 77–79

omnibus test, 138
one-sample *t*-test, 85–88
 alternatives to, 88
 degrees of freedom, 86
 examples in the literature, 88
outcome variable, 39
outliers, 70, 125–126

p value, 62–64, 80–81
paired samples *t*-test, *see* dependent samples *t*-test
parameters, 11
parametric statistical test, 70–72

path analysis, 7–8, 163–185
AMOS, calculate using, 182–185
assumptions, 176–177
decomposing correlation, 166–167
estimating path model, 178–180
examples in literature, 180–181
model identification, 177–178
nonrecursive model, 168
path coefficient, 168, 172–173
path decomposition, 168, 175–176
path diagram, 165, 168–172
path equations, 173–175
recursive model, 168
relationships between variables, 171–172, 175
percentage distributions, 18–20
cumulative percentage distributions, 20
using grouped data, 20
percentile, 30, 46–49
pie charts, 20–21
planned contrasts or comparisons, 144–148, 161
nonorthogonal, 145–146
orthogonal contrasts, 145–147
Planned Orthogonal Contrast procedure, SPSS, 147
SPSS, calculate using, 161
weights, 145–146
posthoc contrasts or comparisons, 148–149, 162–163
SPSS, calculate using 162–163
power, 65, 69
predictor variable, 39

random measurement error, 14, 53
random assignment, 50, 54
random sampling, 51–52
range, 29–30, 32–33
interquartile range, 30
maximum, 29–30
minimum, 29–30
SPSS, calculate using, 32–33
rejection level, 60–61, 63–64, 81
reliability of measurement, 14
replication, 54
research hypotheses, 36–37, 61–62, 69
directional hypothesis, 37, 61–62
nondirectional hypothesis, 37, 61–62
one-tailed hypothesis, 37, 61–62
two-tailed hypothesis, 37, 61–62
research sampling, 50–54
nonrandom sampling, 51–52
random sampling, 51–52
sampling bias, 53–54
sampling error, 54
residuals, 115–116, 136, 143, 155, 158, 177
error, 136
plots, 143
residual variable, 168–170

Sample Power, 126
sampling distribution of the differences between means, 58–60
sampling distribution of means, 55–56
scattergram, 74–75, 101, 107–108
scatterplot, 115–117, 126
Scheffe, 149
shapes of data distributions, 41–49
kurtosis, 41–42, 49
normal distribution, 42–49
skewness, 41–43, 49
significance level, *see* rejection level
significance testing, 62–64
skewness, 41–43, 49, 70
positively skewed distributions, 41–42
negatively skewed distributions, 41–43
slope, 111, 113–114, 121
Spearman's rho, 108
spread, 29
SPSS, calculate using, 28–29, 32–33, 83–84, 93–94, 98–99, 106–107, 129–130, 147, 160–162
ANOVA, 160–162
chi-square test of association, 83–84
correlation, 106–107
dependent samples *t*-test, 93–94
homogeneity of variance, 129–130
independent samples *t*-test, 98–99
mean, 28–29
measures of central tendency, 28–29
measures of variability, 32–33
median, 28–29
mode, 28–29
multiple linear regression analysis, 129–130
planned contrasts or comparisons, 161
Planned Orthogonal Contrast procedure, 147
post-hoc contrasts or comparisons, 161–162

range, 32–33
standard deviation, 32–33
variance, 32–33
squared multiple correlation coefficient, 119
standard deviation, 29, 32–33
SPSS, calculate using, 32–33
standard score, *see z* score
standard error of the differences between means, 59–60
standard error of the mean, 56
statistical levels of measurement, 17–18
dichotomous, 18
scale level, 18
statistical test, selecting the appropriate, 69
statistic, 11
structural equation modeling, 164, 181–182
measurement model, 182
structural model, 182
sum of squared differences, 31, 116–119, 139
model sum of squared differences, 116–119
residual sum of squared difference, 116–119
total sum of squared differences, 116–119
sum of squares, 154
between-groups sum of squares, 154
within-groups sum of squares, 154
suppressor variable, 40
systematic measurement error, 14, 52–53

t-test, 85–99, 121, 133
bridge to advanced statistical tests, 99, 133
dependent samples, 88–94
evaluating estimated coefficients, 121
independent samples, 94–99
one-sample, 85–88
Tamabane's T2, 149
temporal ordering, 170
Tukey-Kramer test, 147
Tukey's Honestly Significant Difference test, 149
Type I error, 64–65, 123, 137–138, 149
family-wise or experiment-wise, 138
Type II error, 64–65, 123, 125

univariate statistical methods, 2, 11

validity of measurement, 14
variability, measures of, 29–33
dispersion, 29
range, 29–30, 32–33
spread, 29
SPSS, calculate using, 32–33
standard deviation, 29, 32–33
variance, 29–33
variable, 12–15, 39–41, 167–170, 181–182
attributes, 13
categorical, 13
conceptualization, 12
confounding, 40–41
continuous, 13 control, 40
dependent, 39
dichotomous, 13, 15
discrete, 13
endogenous, 167, 169
exogenous, 167
independent, 39
intervening, 40, 167
latent, 181–182
mediator, 39–40, 168
moderator, 39–40
obscuring, 40
operationalization, 12
outcome, 39
predictor, 39
residual, 168–170
suppressor, 40
variance, 29–33, 70–71, 119–121, 123–125, 129–130, 136, 138–140, 143, 145, 157–158
between-groups, 139–140, 145
heteroscedasticity, 70–71, 123–124
homogeneity of, 70–71, 123–124, 129–130, 143
homoscedasticity, *see* homogeneity of variance
inflation factor, 125
partitioning, 157–158
proportion of, 119–121
SPSS, calculate using, 32–33
within-groups, 139–140, 145
variance inflation factor, 125

weighted least-square method, 71

z score, 46–49, 67

CPSIA information can be obtained
at www.ICGtesting.com
Printed in the USA
BVHW040251290821
615364BV00017B/216

9 780199 764044